D1663932

Electronic Properties of Metals

Materials Science Monographs, 64

Electronic Properties of Metals

Gerd Lehmann

Zentralinstitut für Festkörperphysik und Werkstofforschung
Akademie der Wissenschaften der DDR
Dresden, G.D.R.

Paul Ziesche

Sektion Physik
Technische Universität Dresden, G.D.R.

ELSEVIER
Amsterdam – Oxford – New York – Tokyo 1990

Elsevier Science Publishers
25 Sara Burgerhartstraat
P. O. Box 211, 1000 AE Amsterdam, The Netherlands

Distributors for the U.S.A. and Canada

Elsevier Science Publishing Company, Inc.
655 Avenue of the Americas
New York, NY 10010

ISBN 0-444-98838-6 (Vol. 64)
ISBN 0-444-41685-4 (Series)

With 63 Figures and 6 Tables

Printed in the German Democratic Republic

MATERIALS SCIENCE MONOGRAPHS (Advisory Editor: C. LAIRD)

Vol. 1 Dynamic Mechanical Analysis of Polymeric Material (Murayama)

Vol. 2 Laboratory Notes on Electrical and Galvanomagnetic Measurements (Wieder)

Vol. 3 Electrodeposition of Metal Powders (Căluşaru)

Vol. 4 Sintering — New Developments (Ristič)

Vol. 5 Defects and Diffusion in Solids. An Introduction (Mrowec)

Vol. 6 Energy and Ceramics (Vincenzini)

Vol. 7 Fatigue of Metallic Materials (Klesnil and Lukáš)

Vol. 8 Synthetic Materials for Electronics (Jakowlew, Szymański and Włosinski)

Vol. 9 Mechanics of Aerospace Materials (Nica)

Vol. 10 Reactivity of Solids (Dyrek, Haber and Nowotny)

Vol. 11 Stone Decay and Conservation (Amoroso and Fassina)

Vol. 12 Metallurgical Aspects of Environmental Failures (Briant)

Vol. 13 The Use of High Intensity Ultrasonics (Puškár)

Vol. 14 Sintering — Theory and Practice (Kolar, Pejovnik and Ristić)

Vol. 15 Transport in Non-Stoichiometric Compounds (Nowotny)

Vol. 16 Ceramic Powders (Vincenzini)

Vol. 17 Ceramics in Surgery (Vincenzini)

Vol. 18 Intergranular Corrosion of Steels and Alloys (Číhal)

Vol. 19 Physics of Solid Dielectrics (Bunget and Popescu)

Vol. 20 The Structure and Properties of Crystal Defects (Paidar and Lejček)

Vol. 21 Interrelations between Processing Structure and Properties of Polymeric Materials (Seferis and Theocaris)

Vol. 22 Atmospheric Deterioration of Technological Materials: A Technoclimatic Atlas (Rychtera)
Part A: Africa
Part B: Asia (excluding Soviet Asia), Australia and Oceania

Vol. 23 Plasma Metallurgy (Dembovský)

Vol. 24 Fatigue in Materials: Cumulative Damage Processes (Puškár and Golovin)

Vol. 25 Sintered Metal-Ceramic Composites (Upadhyaya)

Vol. 26 Frontiers in Materials Technologies (Meyers and Inal)

Vol. 27 Joints with Fillet Welds (Faltus)

Vol. 28 Reactivity of Solids (Barret and Dufour)

Vol. 29 Progress in Advanced Materials and Processes: Durability, Reliability and Quality Control (Bartelds and Schliekelmann)

Vol. 30 Non-Ferrous Metals and Alloys (Sedláček)

Vol. 31 Defect Recognition and Image Processing in III—V Compounds (Fillard)

Vol. 32 The Si—SiO_2 System (Balk)

Vol. 33 Perspectives on Biomaterials (Lin and Chao)

Vol. 34 Silicon Nitride in Electronics (Belyi et al.)

Vol. 35 High Tech — The Way into the Nineties (Brunsch, Gölden and Herkert)

Vol. 36 Composite Systems from Natural and Synthetic Polymers (Salmén, de Ruvo, Seferis and Stark)

Vol. 37 Copper Indium Diselenide for Photovoltaic Applications (Coutts, Kazmerski and Wagner)

Vol. 38 High Tech Ceramics (Vincenzini)

Vol. 39 Ceramics in Clinical Applications (Vincenzini)

Vol. 40 Electron Microscopy in Solid State Physics (Bethge and Heydenreich)

Vol. 41 Looking Ahead for Materials and Processes (De Bossu, Briens and Lissac)

Vol. 42 Material Data for Cyclic Loading. Parts A—E (Boller and Seeger)

Vol. 43 Technical Mineralogy and Petrography. Parts A and B (Szymański)

Vol. 44 Defect Recognition and Image Processing in III—V Compounds II (Weber)

Vol. 45 Solid State Electrochemistry and its Applications to Sensors and Electronic Devices (Goto)

Vol. 46 Basic Mechanisms in Fatigue of Metals (Lukáš and Polák)

Vol. 47 Surface and Near-Surface Chemistry of Oxide Materials (Nowotny and Dufour)

Vol. 48 Creep in Metallic Materials (Čadek)

Vol. 49 Hardness Estimation of Minerals, Rocks and Ceramic Materials (A. Szymański and J. M. Szymański)

Continued on page 197

Preface

Though the key to a basic and qualitative understanding of the properties of solids on a microscopic level had already been given by the quantum theory, no more than 20 years ago a stormy development began concerning their detailed quantitative understanding. Particularly the investigation of the electronic structure led to a deep insight into the behaviour of condensed matter.

Strictly speaking the excitation spectrum or the band structure of electrons is often understood as electronic structure. This definition is based on the Landau theory of elementary excitations, which shows that the reaction of a many-particle system on a weak external perturbation can be described by nearly non-interacting low-energy excitations of one-particle type. In metals these excitations near the Fermi energy are damped only very weakly. On that basis many electronic properties, especially of metals, can be understood and increasingly can also be calculated. With this a growing contribution to materials science is made by solid state physics.

Because of limited space, the discussion of the electronic properties of metals can by no means be exhaustive. Therefore, for more detailed studies we refer to a number of good presentations of the subject (Ziman [74], Abrikosov [72] and [87], Kittel [73], Lifshitz et al. [75], Brauer [72], Weissmantel and Hamann [79]; for further references see the register of literature).

In this book we draw attention both to understanding the basic principles of solid state physics and also, and especially, to those actual problems and recent applications where summarizing presentations do not yet exist. Therefore, the depth of treatment of the individual parts and chapters is different.

In Chapter 1 the important statements of the one-particle picture of electrons are formulated including its foundation on the many-particle theory. The usual methods for calculating the electronic structure of ordered and disordered metals by solving the multi-centre problem are explained in Chapter 2. In most cases it is only possible to discuss the physical ideas. A more detailed survey of the results obtained is given in Chapter 3. Whereas Chapter 4 connects the theory with the most important experiments, in Chapter 5 the calculation of macroscopic properties from the electronic structure is explained. In most cases the presentation can only be considered to give a short introduction. For more detailed studies the reader is referred to the quoted reviews and the original literature. Finally, Chapter 6 concerns a recent field of increasing interest, i.e. surfaces, interfaces, and low-dimensionally structured solids.

In preparing the English edition we have included besides new chapters on norm-conserving pseudo-potentials and high temperature superconductors, short sections on the most recent problems, results, and developments together with the corresponding references to the literature. Nevertheless, there are other rapidly developing fields which could not be treated for reasons of space: mixed valence systems, heavy Fermion systems, quantum Hall effect, heterostructures, quantum microstructures, the scanning tunnel and the atomic force microscope, small particles, amorphous metals, solution of the quasi particle equation (which allows the gap-problem in semiconductor theory to be solved and is also of interest for the theory of metals), combination of density-functional theory and molecular dynamics. We would like to mention that a common electron theory for all types of solids is developing, unifying more and more the viewing and treatment of the electronic structure and electronic properties of metals and semiconductors. In this connection also the gradually arising concepts of quantum engineering, band structure engineering, band gap engineering should be mentioned too, indicating the increasing transfer of results of the electron theory to materials science and thus contributing to the theoretical fundamentals of that discipline (see e.g. Weinberger [90], Ehrenreich [87], Pettifor [87], Faulkner [82], Fradin [81]). Some recent monographs and proceedings dealing with electron theory are especially given within the references (in addition to the list in the 1984 German edition).

Discussions with our colleagues R. Lenk, W. John and H. Eschrig during the preparation of the manuscript are gratefully acknowledged. Furthermore we thank J. Gräfenstein, I. Mertig, and E. Mrosan for taking part in the translation as well as our wives B. Lehmann and I. Ziesche for typing the English manuscript and for their patience. We are also bound in gratitude to the publishers for good cooperation.

G. Lehmann
P. Ziesche

Contents

1 One-Particle Picture of Metal Electrons

1.1 Electron Gas

1.1.1 Free Electrons and Metallic Properties

The hypothesis of the existence of free electrons in metals allowed the qualitative interpretation of a series of typically metallic properties and was proved experimentally soon after its formulation by Drude (1900).

Metallic properties include

- high electrical conductivity σ_{el} increasing with decreasing temperature,
- similar high thermal conductivity σ_{th} connected with the electrical conductivity through the Wiedemann-Franz law

$$\sigma_{th} = \frac{\pi^2}{3} \frac{k_B^2}{e^2} T\sigma_{el},$$

- metallic lustre induced by total reflection of light and other optical properties (frequency dependence of optical absorption),
- thermal emission of electrons,
- the photoelectric effect,
- the Hall effect and further thermoelectric and galvanomagnetic effects, and
- the result of Tolman's experiment as a direct indication of the existence of free electrons.

First attempts to explain these properties were of a qualitative nature, and calculations yielded the wrong order for the effects. The clearest evidence for this fact was obtained by measurement of the electronic contribution to the specific heat capacity, which showed that only a small fraction of electrons participate in thermal equilibrium.

The explanation for this was not provided until the appearance of quantum theory and quantum statistics, respectively, for the Fermi gas of free electrons. This will be discussed in more detail in [5.1.1]. Further discrepancies in the investigation of the Hall effect could similarly be explained by quantum theory, if the motion of electrons in a periodic crystal potential is taken into account.

In the simplest picture, the electrons are assumed to move freely in a constant potential. They are prevented from leaving the crystal by electrostatic forces and move in a potential well (fig. 1.1-1). According to Fermi-Dirac statistics (see eqn. (7)), at a temperature $T = 0$ the electrons fill the potential well up to the Fermi energy ε_F. The average occupation number for states below ε_F is equal to one, whereas for states with higher energy it disappears.

At finite temperatures, the occupation of states near ε_F is smeared out over an interval of width $k_B T$. By these statistics, Pauli was also able to explain the temperature-independent spin paramagnetism of electrons (see [5.1.4]).

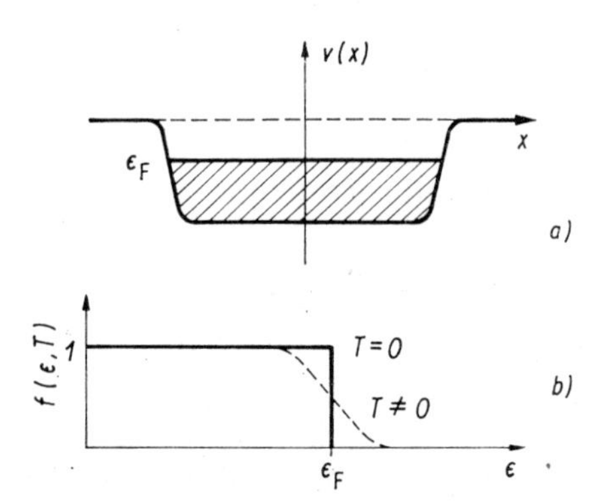

Fig. 1.1-1. a) Schematic representation of the behaviour of the potential $v(x)$ for free electrons in a metal
b) Occupation number of states as a function of energy for zero (solid line) and non-zero (broken line) temperatures

1.1.2 Fermi Sphere

The consideration of electrons in the potential well of fig. 1.1-1a) is not well suited for mathematical treatment. Since the exact shape of the potential near the well surface influences the motion of the electrons only in a very small region (of the order of a few atomic distances), the "surface influence" may be neglected for systems large enough and a more convenient form of boundary conditions may be applied, the so-called periodic volume. Under these conditions, the state of free electrons is determined by their momentum $\boldsymbol{p} = \hbar \boldsymbol{k}$ and spin s. The wave functions are described by plane waves $\psi(\boldsymbol{r}) \sim \exp(i\boldsymbol{k}\boldsymbol{r})$ and their energy is given by $\varepsilon = p^2/2m$. The periodic volume means that the wave function repeats in the direction of the coordinate axis (here for the x-direction) after the length L_x of the periodic volume:

$$\psi(x + L_x) = \psi(x) . \tag{1}$$

This is equivalent to the condition

$$k_x L_x = 2\pi n_x \quad \text{with} \quad n_x = 0, \pm 1, \pm 2, \ldots \tag{2}$$

for the wave number k_x in the x-direction.

Thus, in three-dimensional $\boldsymbol{k}$-space, the allowed states are given by a point lattice with edge length $\Delta k_i = 2\pi/L_i$ ($i = x, y, z$), see fig. 1.1-2. As a consequence of the Pauli principle, each state may be doubly occupied as two spin states exist for each electron. Therefore, in the ground state of the total system, the N electrons occupy the states inside the Fermi sphere with radius $k_F = \sqrt{2m\varepsilon_F}/\hbar$ as shown in fig. 1.1-2. The total energy of the system is given by

$$E = \sum_{\boldsymbol{k}}^{\text{occ}} (\hbar \boldsymbol{k})^2/2m\,. \tag{3}$$

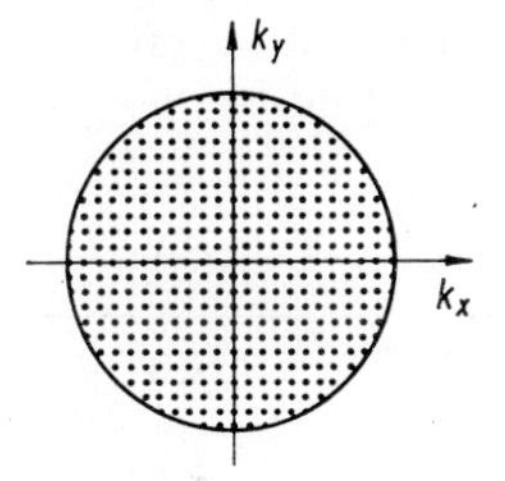

Fig. 1.1-2. Allowed one-particle states in $\boldsymbol{k}$-space. In the ground state of the total system the one-particle states below the Fermi energy are occupied (Fermi sphere)

For very large particle numbers N, the sum in (3) over the occupied states can be replaced by an integral over $\boldsymbol{k}$-space using (2):

$$\sum_{\boldsymbol{k}}^{\text{occ}} \to 2 \int \mathrm{d}n_x\, \mathrm{d}n_y\, \mathrm{d}n_z = \frac{2\Omega}{(2\pi)^3} \int \mathrm{d}^3k\,. \tag{4}$$

The volume element in $\boldsymbol{k}$-space $\Delta k_x\, \Delta k_y\, \Delta k_z = (2\pi)^3/L_x L_y L_z$ or $\mathrm{d}^3k = (2\pi)^3/\Omega$ can be occupied by two electrons inducing the factor 2 in (4). $\Omega = L_x L_y L_z$ denotes the volume of the periodic region. The particle number N, their density n, and energy E are given by

$$N = \frac{2\Omega}{(2\pi)^3} \frac{4\pi}{3} k_F^3 \quad \text{or} \quad n = \frac{N}{\Omega} = \frac{1}{3\pi^2} k_F^3 \tag{5}$$

and

$$E = \frac{2\Omega}{(2\pi)^3} \int_{k \leq k_F} \mathrm{d}^3k \frac{\hbar^2 k^2}{2m} = \frac{3}{5} N \varepsilon_F \quad \text{or} \quad \bar{\varepsilon} = \frac{E}{N} = \frac{3}{5} \varepsilon_F\,. \tag{6}$$

(5) gives the connection between particle density n and Fermi wave number k_F. Because of the high kinetic energy in the small available volume, systems with large density n have a total energy per volume increasing as $n^{5/3}$.

For finite temperatures, one has to take into account the temperature-dependent occupation of the states in (5) and (6) according to Fermi-Dirac statistics

$$f(\varepsilon, T) = (\mathrm{e}^{(\varepsilon - \zeta)/k_B T} + 1)^{-1} \tag{7}$$

(see fig. 1.1-1b)). In the main, it represents a smearing out of width $k_B T$ near the chemical potential $\zeta(T)$, whereas $\zeta(T)$ determines the average particle density of the system at $T \neq 0$. For zero temperature, it equals the Fermi energy: $\zeta(0) = \varepsilon_F$. Since for metals the Fermi energy is of the order of 10 eV (Cu: 7 eV) as a rule $k_B T$ is small compared with ε_F, for the energy of $k_B T = 1$ eV corresponds to a temperature of 11600 K! Because of the condition $k_B T \ll \varepsilon_F$, this is called a degenerate electron gas. The temperature influence induces in (5) and (6) only small correction terms proportional to T^2. This means that only a few of the free electrons present participate in thermal equilibrium and transport processes. A more detailed investigation of these effects will be given in [4.1] and [5.1]. Assuming one free electron per atom for monovalent metals, the following values for density n are obtained

	Li	Na	Cu	Ag
$n/10^{28}\ \mathrm{m}^{-3}$	4.6	2.5	8.5	5.6
$r_s = \left(\frac{3}{4\pi n}\right)^{1/3} a_0^{-1}$	3.2	3.9	2.67	3.06

r_s is the usual dimensionless density parameter of a homogeneous electron gas.

1.2 Band Structure

1.2.1 Bloch Condition

The appearance of a band structure with the characteristic feature of energy regions allowed or forbidden for the motion of electrons is a consequence of the periodic structure of the crystal. To make this fact more transparent, first we investigate the motion of an electron in an external potential $v(\boldsymbol{r})$ exhibiting crystal symmetry. The electron is described by the stationary Schroedinger equation

$$\left(-\frac{\hbar^2}{2m}\frac{\partial^2}{\partial \boldsymbol{r}^2} + v(\boldsymbol{r})\right)\psi(\boldsymbol{r}) = \psi(\boldsymbol{r})\,\varepsilon\,. \tag{1}$$

This partial differential equation of second order yields the time-independent wave function $\psi(\boldsymbol{r})$ and the energy eigenvalues ε both containing all the information necessary for description of the properties and behaviour of the electron. The eigenvalues are determined by the condition that the absolute value $|\psi|^2$ of the wave function may be interpreted physically as the density for the probability of finding the electron at position $\boldsymbol{r}$ and, therefore, that it is normalizable in the available total volume

$$\int_\Omega \mathrm{d}^3r\, |\psi(\boldsymbol{r})|^2 = 1\,. \tag{2}$$

The crystal symmetry of the potential is expressed by

$$v(\boldsymbol{r} + \boldsymbol{R}) = v(\boldsymbol{r})\,. \tag{3}$$

This means that the potential is invariant against a shift by a lattice vector $\boldsymbol{R}$. An arbitrary lattice vector $\boldsymbol{R}$ may be represented by a sum of multiples of the three basic lattice vectors $\boldsymbol{a}_i$:

$$\boldsymbol{R} = m_1\boldsymbol{a}_1 + m_2\boldsymbol{a}_2 + m_3\boldsymbol{a}_3 . \tag{4}$$

m_1, m_2, and m_3 are integer numbers. The so-called Bravais lattice of the total crystal may be constructed by a repeated (N-fold) shift of an unit (or elementary cell) by each allowed lattice vector (4) (see Paufler/Leuschner 1975). The volume of the elementary cell is $\Omega_0 = \Omega/N$ with N as the number of elementary cells in the crystal.

The planes perpendicular to the vectors $\boldsymbol{R}$ to the neighbouring points of a considered lattice point, and cutting them at their midpoints, limit the so-called Wigner-Seitz cell (WS cell). A sphere of the same volume is called a WS sphere. In the case of primitive lattices, i.e. lattices with only one atom per unit cell, the WS cell is identical to the elementary cell.

The translational symmetry described by (3) simplifies considerably the solution of the Schroedinger equation (1), as the periodicity of the crystal potential is expressed in the wave functions in the form of the Bloch theorem:

$$\psi_{\boldsymbol{k}\nu}(\boldsymbol{r} + \boldsymbol{R}) = e^{i\boldsymbol{k}\boldsymbol{R}}\psi_{\boldsymbol{k}\nu}(\boldsymbol{r}) . \tag{5}$$

According to (5), the wave function is not invariant as the potential (3) but will be multiplied by a constant factor $\exp(i\boldsymbol{k}\boldsymbol{R})$, if it is shifted by a lattice vector $\boldsymbol{R}$ from one cell to another. Because of the normalization condition (2), this factor must be of modulus one. However, for finite systems other values are also possible, describing exponentially increasing or decreasing solutions. They may characterize states near surfaces or interfaces, for example see [6.1].

Of course, the probability density $|\psi|^2$ as an observable quantity according to (5) is invariant just as the potential. Other possibilities for the representation of (5) are

$$\psi_{\boldsymbol{k}\nu}(\boldsymbol{r}) = N^{-1/2}\, e^{i\boldsymbol{k}\boldsymbol{r}} u_{\boldsymbol{k}\nu}(\boldsymbol{r}) \quad \text{with} \quad u_{\boldsymbol{k}\nu}(\boldsymbol{r} + \boldsymbol{R}) = u_{\boldsymbol{k}\nu}(\boldsymbol{r}) \tag{5'}$$

and (for primitive lattices)

$$\psi_{\boldsymbol{k}\nu}(\boldsymbol{r}) = N^{-1/2} \sum_{\boldsymbol{R}} e^{i\boldsymbol{k}\boldsymbol{R}} w_\nu(\boldsymbol{r} - \boldsymbol{R}) . \tag{5''}$$

The normalization integral (2) yields for each of the N elementary cells of volume Ω_0 the same value and, therefore, can be written as

$$\int_{\Omega} d^3r\, |\psi_{\boldsymbol{k}\nu}(\boldsymbol{r})|^2 = \int_{\Omega_0} d^3r\, |u_{\boldsymbol{k}\nu}(\boldsymbol{r})|^2 = 1 . \tag{2'}$$

(2′) demonstrates why the normalization factor N appears in (5′) and (5″).

In each representation, $\boldsymbol{k}$ characterizes the eigenvalue of the translational operator changing the argument $\boldsymbol{r}$ of the wave function by a lattice vector $\boldsymbol{R}$. The band index ν discriminates different solutions of the Schroedinger equation (1) with the same translational properties (5). The first factor in (5') gives an indication for the similarity of $\boldsymbol{k}$ with the momentum $\boldsymbol{p} = \hbar\boldsymbol{k}$ of a free particle which is represented by a plane wave $\exp(i\boldsymbol{k}\boldsymbol{r})$. Because of this relation, $\hbar\boldsymbol{k}$ is denoted as quasi-momentum. However, $\boldsymbol{k}$ in (5) is only determined up to vectors $\boldsymbol{G}$ of the reciprocal lattice which are defined by relation (7) because $\exp(i\boldsymbol{k}\boldsymbol{R})$ and $\exp\left(i(\boldsymbol{k} + \boldsymbol{G})\,\boldsymbol{R}\right)$ are identical eigenvalues of the translational operator[1]) as a consequence of $\exp(i\boldsymbol{G}\boldsymbol{R}) = 1$ if the condition

$$\boldsymbol{G}\boldsymbol{R} = 2\pi s \quad (s \text{ integer}) \tag{6}$$

is fulfilled. In analogy to (4), $\boldsymbol{G}$ may be represented by the basic vectors $\boldsymbol{b}_j$ of the reciprocal lattice:

$$\boldsymbol{G} = n_1\boldsymbol{b}_1 + n_2\boldsymbol{b}_2 + n_3\boldsymbol{b}_3\,. \tag{7}$$

The $\boldsymbol{a}_i$ and $\boldsymbol{b}_j$ are connected by the relation

$$\boldsymbol{a}_i\boldsymbol{b}_j = 2\pi\delta_{ij} \tag{8}$$

(see fig. 1.2-1).
It is evident that the relation (6) is fulfilled by means of (7) and (8) for arbitrary values of $\boldsymbol{G}$ and $\boldsymbol{R}$.

The representation (5') is well suited to the description of nearly free electrons. Then the periodic function $u_{\boldsymbol{k}\nu}(\boldsymbol{r})$ characterizes the deviations of the

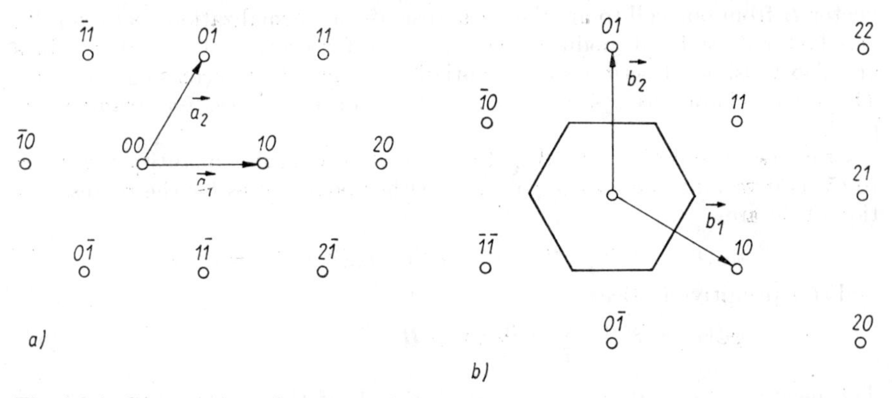

Fig. 1.2-1. Construction of vectors of the reciprocal lattice (two-dimensional)
a) Lattice vectors of real space
b) Vectors of the reciprocal lattice

[1]) We refer to the fact that as opposed to free particles, for lattice electrons the wave vector $\boldsymbol{k}$ has only a physical meaning in the combination $\exp(i\boldsymbol{k}\boldsymbol{R})$. Here $\boldsymbol{R}$ denotes a finite lattice vector. For free electrons, the potential is constant, i.e. invariant against translation by arbitrary vectors $\boldsymbol{R}$. Therefore, for free electrons each $\boldsymbol{k}$ denotes a different physical state.

crystal electrons from the behaviour of free electrons (then u differs only insignificantly from unity). On the other hand, (5″) corresponds more to the behaviour of tightly bound electrons. The wave function $\psi_{\boldsymbol{k}\nu}(\boldsymbol{r})$ in (5″) is generated by superposition of the so-called Wannier functions $w_\nu(\boldsymbol{r})$ which are localized at the given ion site $\boldsymbol{R}$ of the lattice. The sum runs over all lattice sites $\boldsymbol{R}$. In the case of tightly bound electrons and large atomic distances (in comparison with extension of the states), the $w_\nu(\boldsymbol{r})$ are identical to the strongly localized atomic wave functions. In this limiting case, the index ν denotes the atomic levels. In general, states $w_\nu(\boldsymbol{r})$ may always be constructed which are extended on a few elementary cells only (compare [2.1]).

Also for lattice electrons, the periodic boundary conditions known from the free electron gas [1.1.2] are used. Of course, for small systems (also possibly in only one direction) the consideration of boundary conditions corresponding to the real physical situation is necessary. In the following, we choose periodic boundary conditions in a form where the periodic volume with shape of a parallelepiped consists of directions $\boldsymbol{a}_i$ of N_i elementary cells, respectively. The possible $\boldsymbol{k}$-values constitute a discrete manifold

$$\boldsymbol{k} = \frac{m_1}{N_1}\boldsymbol{b}_1 + \frac{m_2}{N_2}\boldsymbol{b}_2 + \frac{m_3}{N_3}\boldsymbol{b}_3 \tag{9}$$

with integers m_i,

$$0 \leqq m_i < N_i .$$

They form a fine periodic lattice. The volume $\Delta^3 k$ between neighbouring points is

$$\Delta^3 k = (N_1 N_2 N_3)^{-1}\,[\boldsymbol{b}_1, \boldsymbol{b}_2, \boldsymbol{b}_3] = (2\pi)^3/N\Omega_0 = (2\pi)^3/\Omega \tag{10}$$

or $\Omega \mathrm{d}^3 k = (2\pi)^3$, respectively.
In (10) the connection between the scalar triple product $[\boldsymbol{b}_1, \boldsymbol{b}_2, \boldsymbol{b}_3]$ of the basis vectors $\boldsymbol{b}_i$ of the reciprocal lattice and volume $\Omega_0 = [\boldsymbol{a}_1, \boldsymbol{a}_2, \boldsymbol{a}_3]$ of the elementary cell

$$[\boldsymbol{b}_1, \boldsymbol{b}_2, \boldsymbol{b}_3] = (2\pi)^3/[\boldsymbol{a}_1, \boldsymbol{a}_2, \boldsymbol{a}_3],$$

following from (8) has been used. The total number of elementary cells $N_1\,N_2\,N_3$ is denoted by N.

Functions $\psi_{\boldsymbol{k}\nu}(\boldsymbol{r})$ are solutions of the Schroedinger equation (1) with the hermitian Hamiltonian and, therefore, are orthogonal over the total volume Ω:

$$\int_\Omega \mathrm{d}^3 r\, \psi^*_{\boldsymbol{k}\nu}(\boldsymbol{r})\, \psi_{\boldsymbol{k}'\nu'}(\boldsymbol{r}) = \delta_{\boldsymbol{k}\boldsymbol{k}'}\delta_{\nu\nu'} . \tag{2''}$$

As regards $\boldsymbol{k}$ and $\boldsymbol{k}'$ which are situated in the elementary cell of $\boldsymbol{k}$-space, the first Brillouin zone (which will be treated in more detail in the next section), the orthogonality (2″) follows from the Bloch theorem, by means of which (2″)

may be reduced on an integral over the elementary cell. The sum over phase factors

$$\sum_{\boldsymbol{R}} e^{i(\boldsymbol{k}'-\boldsymbol{k})\boldsymbol{R}} = N \sum_{\boldsymbol{G}} \delta_{\boldsymbol{k}-\boldsymbol{k}',\boldsymbol{G}} \tag{11}$$

created by this operation gives exactly the Kronecker symbol with respect to $\boldsymbol{k}$ and $\boldsymbol{k}'$. The terms in (11) cancel out by interference, if $\boldsymbol{k}$ and $\boldsymbol{k}'$ are not equal or differ only by a reciprocal lattice vector $\boldsymbol{G}$, respectively. In the latter case because of (6), all terms equal unity. As has been presumed, the different solutions $\psi_{\boldsymbol{k}\nu}(\boldsymbol{r})$ to the same $\boldsymbol{k}$ are orthogonal in the elementary cell, leading to the Kronecker $\delta_{\nu\nu'}$ in (2″). It may be shown that also the Wannier functions in (5″) fulfil the orthonormality relation

$$\int d^3r w_\nu(\boldsymbol{r}-\boldsymbol{R})\, w_{\nu'}(\boldsymbol{r}-\boldsymbol{R}') = \delta_{\nu\nu'}\delta_{\boldsymbol{R}\boldsymbol{R}'}\,. \tag{2‴}$$

1.2.2 Energy Bands, Brillouin Zones, Fermi Surfaces, Density of States

In analogy to (5), the energy eigenvalues of (1) like $\psi(\boldsymbol{r})$ depend both on $\boldsymbol{k}$ and ν:

$$\varepsilon = \varepsilon_\nu(\boldsymbol{k}) \quad \text{with} \quad \varepsilon_\nu(\boldsymbol{k}+\boldsymbol{G}) = \varepsilon_\nu(\boldsymbol{k})\,. \tag{12}$$

Because of the physical identity (6), the energies for states $\boldsymbol{k}$ and $\boldsymbol{k}+\boldsymbol{G}$ must be identical, meaning that $\varepsilon(\boldsymbol{k})$ is a function periodic in $\boldsymbol{k}$-space whose periodicity is given by the reciprocal lattice.

A further property of $\varepsilon_\nu(\boldsymbol{k})$ which should be real as an eigenvalue of (1) follows, if we take the conjugated complex of (1):

$$\left(-\frac{\hbar^2}{2m}\frac{\partial^2}{\partial \boldsymbol{r}^2} + v(\boldsymbol{r})\right)\psi^*_{\boldsymbol{k}\nu}(\boldsymbol{r}) = \psi^*_{\boldsymbol{k}\nu}(\boldsymbol{r})\,\varepsilon_\nu(\boldsymbol{k})\,. \tag{1'}$$

According to (5), the function $\psi^*_{\boldsymbol{k}\nu}(\boldsymbol{r})$ fulfils the condition

$$\psi^*_{\boldsymbol{k}\nu}(\boldsymbol{r}+\boldsymbol{R}) = e^{-i\boldsymbol{k}\boldsymbol{R}}\psi^*_{\boldsymbol{k}\nu}(\boldsymbol{r})\,.$$

This is identical to the Bloch condition for a function $\psi_{-\boldsymbol{k}\nu}(\boldsymbol{r})$. If only one eigenfunction exists for any value $\boldsymbol{k}$, then

$$\psi_{-\boldsymbol{k}\nu}(\boldsymbol{r}) = \psi^*_{\boldsymbol{k}\nu}(\boldsymbol{r}) \tag{13}$$

holds. From (1′) and (13) we immediately obtain

$$\varepsilon_\nu(\boldsymbol{k}) = \varepsilon_\nu(-\boldsymbol{k})\,. \tag{14}$$

Consequently, the energy represents a function in $\boldsymbol{k}$-space symmetrical against inversion, as already known from free particles ($\varepsilon = \hbar^2\boldsymbol{k}^2/2m$). This symmetry is connected with the invariance against time inversion.

Before calculation of the energy eigenvalues, we wish to discuss the motion in a periodic potential by the example of the one-dimensional case. In fig. 1.2-2 in periodic potential $v(x)$, states with energies ε_1 and ε_2 are registered. In classical physics, a state with energy ε_1 represents a bound particle, whereas a state with energy ε_2 concerns a free one. The two states are different.

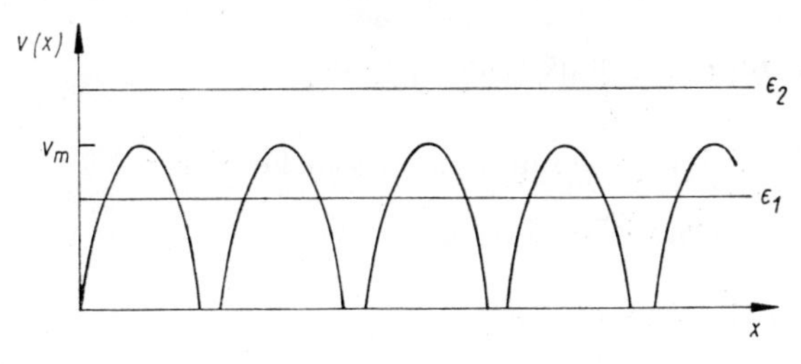

Fig. 1.2-2. Motion in a periodic potential

In quantum mechanics the difference is not so pronounced: also for energies $\varepsilon_2 < v_m$ (maximum of the potential), the wave packets are reflected at the periodic potential causing a behaviour very different from classical if the waves reflected at the different potential wells are "in phase" (therefore, for certain wavelengths and energies wave propagation is impossible, these energy regions are called "forbidden" regions or gaps). On the other hand, for $\varepsilon_1 < v_m$ the electrons may penetrate the potential maxima because of the tunnel effect and, therefore, obtain a certain residual mobility.

The appearance of forbidden gaps is typical for periodic potentials of arbitrary shape. The energy gaps separate regions with a continuous (in the limiting case $\Omega \to \infty$) spectrum of eigenvalues $\varepsilon_\nu(\boldsymbol{k})$, the so-called bands, where the electrons may move through the crystal. This motion is not possible within the forbidden gaps.

For a better understanding, we still investigate two limiting cases:

- the approximation starting from free electrons
- the approximation starting from free atoms.

For free electrons in a constant potential $v = 0$, we have

$$\varepsilon^{(0)}(\boldsymbol{k}) = \hbar^2\boldsymbol{k}^2/2m\,, \quad \psi_{\boldsymbol{k}}^{(0)}(\boldsymbol{r}) = \Omega^{-1/2}\,\mathrm{e}^{i\boldsymbol{k}\boldsymbol{r}}. \tag{15}$$

The influence of a periodic potential assumed to be small is described by Schroedinger perturbation theory, where the free electron appears as the zero order (therefore we write $\varepsilon^{(0)}$ and $\psi^{(0)}$ in (15)). The first order $\varepsilon^{(1)}$ simply gives a shift of energy by the average potential

$$\Delta\varepsilon^{(1)}(\boldsymbol{k}) = \frac{1}{\Omega}\int \mathrm{d}^3r v(\boldsymbol{r}) = \bar{v} \tag{16}$$

independent of $\boldsymbol{k}$. In second order, a more complicated expression

$$\Delta\varepsilon^{(2)}(\boldsymbol{k}) = \sum_{\boldsymbol{k}'(\neq\boldsymbol{k})} \frac{|\langle\boldsymbol{k}|\,v\,|\boldsymbol{k}'\rangle|^2}{\varepsilon^{(0)}(\boldsymbol{k}') - \varepsilon^{(0)}(\boldsymbol{k})} \tag{17}$$

is obtained. $\langle\boldsymbol{k}|\,v\,|\boldsymbol{k}'\rangle$ denotes the matrix element of the potential between plane waves with wave number vectors $\boldsymbol{k}$ and $\boldsymbol{k}'$

$$\langle\boldsymbol{k}|\,v\,|\boldsymbol{k}'\rangle = \frac{1}{\Omega}\int \mathrm{d}^3r\;\mathrm{e}^{-i\boldsymbol{k}\boldsymbol{r}} v(\boldsymbol{r})\,\mathrm{e}^{i\boldsymbol{k}'\boldsymbol{r}}. \tag{18}$$

The periodic potential may be represented as a Fourier series

$$v(\boldsymbol{r}) = \sum_{\boldsymbol{G}} v_{\boldsymbol{G}}\,\mathrm{e}^{i\boldsymbol{G}\boldsymbol{r}}. \tag{19}$$

Therefore, the matrix elements

$$\langle\boldsymbol{k}|\,v\,|\boldsymbol{k}'\rangle = \sum_{\boldsymbol{G}} v_{\boldsymbol{G}}\delta_{\boldsymbol{k}-\boldsymbol{k}',\boldsymbol{G}} \tag{20}$$

are different from zero, only if the difference of $\boldsymbol{k}$ and $\boldsymbol{k}'$ equals a vector $\boldsymbol{G}$ of the reciprocal lattice as expressed by the Kronecker symbol $\delta_{\boldsymbol{k}-\boldsymbol{k}',\boldsymbol{G}}$ in (20). The strength of the transition from the plane wave $\exp(i\boldsymbol{k}\boldsymbol{r})$ to the plane wave $\exp\left(i(\boldsymbol{k}+\boldsymbol{G})\,\boldsymbol{r}\right)$ induced by the potential is given by the Fourier coefficient $v_{\boldsymbol{G}}$,

$$v_{\boldsymbol{G}} = \frac{1}{\Omega}\int \mathrm{d}^3r\;\mathrm{e}^{-i\boldsymbol{G}\boldsymbol{r}} v(\boldsymbol{r}), \tag{21}$$

of the potential.

Because of the smallness of the potential, the contribution of second order (17) is important only if the energy denominator nearly disappears, i.e. if, according to (15), $\boldsymbol{k}$ is near the values given by

$$k^2 = (\boldsymbol{k} - \boldsymbol{G})^2 \quad \text{or} \quad 2\boldsymbol{k}\boldsymbol{G} = G^2. \tag{22}$$

In that case, the perturbation theory is invalid and the equal eigenvalues $\varepsilon^{(0)}(\boldsymbol{k})$ and $\varepsilon^{(0)}(\boldsymbol{k}+\boldsymbol{G})$ split by a quantity $2v_{\boldsymbol{G}}$. This is shown in fig. 1.2-3 for the one-dimensional case. According to (22) deviations from parabolic behaviour of free electrons appear at values $k = G/2$. In fig. 1.2-3b) the periodicity property (12) has been used. Whereas the representation of fig. 1.2-3a) is suitable for describing the behaviour of free electrons and $\varepsilon(k)$ is defined as a function over the whole k-space but with discontinuities, we need in the representation of fig. 1.2-3b) only a knowledge of the continuous but multivalued function $\varepsilon_\nu(k)$ in the basic region from $-G/2$ to $G/2$ (the first Brillouin zone). In this representation, the bands separated by forbidden regions of width $2v_{\boldsymbol{G}}$ are evident.

The appearance of single bands will become clearer considering the opposite limiting case — the approximation from free atoms. For this, we assume the N atoms of the crystal to be arranged in a regular lattice with a lattice constant a differing in an gedanken-experiment from the observed on a_{exp}. In the limit $a \to \infty$, we have non-interacting atoms, and observe their discrete eigenvalues occasionally degenerated N-fold. Reducing a, this degeneration is lifted as a consequence of the overlap of neighbouring wave functions: The isolated degenerate levels broaden into bands (fig. 1.2-4). The broadening at

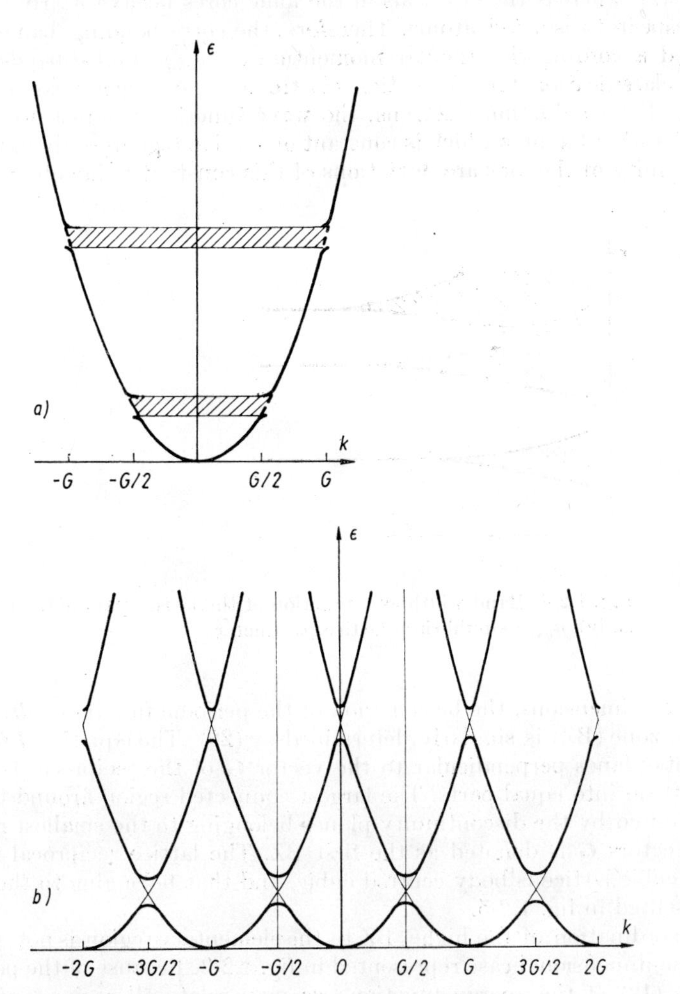

Fig. 1.2-3. a) Dispersion relation of electrons in a one-dimensional potential (solid line) and for free electrons (broken line), respectively. Hatched region: forbidden region
b) Dispersion relation in the periodic scheme

the equilibrium value a_{exp} is small for tightly bound and localized electrons of the ionic core, as in this case the neighbouring atom represents only a small perturbation. For the weakly bound and more extended valence electrons (e.g. in alkaline metals), perturbation by the neighbours is strong, so that broad bands result which may also overlap and approximate to the limit of free electrons. Consequently, the conduction electrons move nearly freely through the crystal, whereas the electrons in the ionic cores behave nearly unchanged with respect to isolated atoms. Therefore, the corresponding bands may be classified according the angular momentum as s-, p- and d-bands, whereas such a classification for conduction electrons is no longer possible unambiguously. For conduction electrons, the wave function has instead the structure (5′) with a factor u which is constant over wide regions (only in the immediate vicinity of the ions are deviations of this constant value observed).

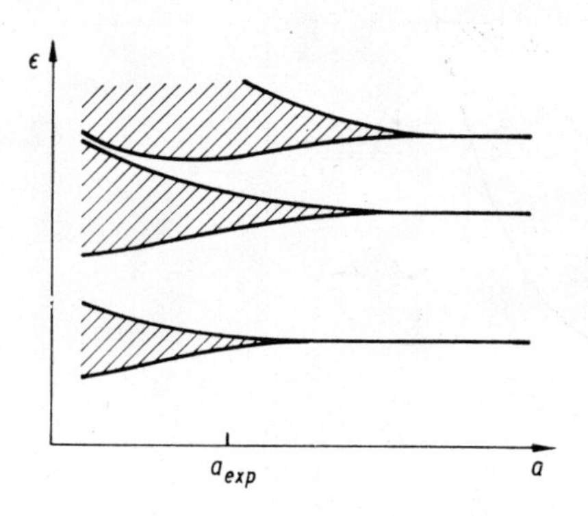

Fig. 1.2-4. Band width as a function of the lattice parameter a (schematically). a_{exp}: equilibrium lattice parameter

In three dimensions, the basic region of the periodic function $\varepsilon_\nu(\boldsymbol{k})$, the first Brillouin zone (BZ) is similarly determined by (22). The equation $\boldsymbol{k}\boldsymbol{G} = \boldsymbol{G}^2/2$ represents planes perpendicular to the vector $\boldsymbol{G}$ of the reciprocal lattice and divides them into equal parts. The largest connected region around the origin $\boldsymbol{k} = 0$ limited by the discontinuity planes belonging to the smallest reciprocal lattice vectors $\boldsymbol{G}$ is denoted as the first BZ. The lattice reciprocal to a face centred cubic lattice is body centred cubic, and that belonging to the first BZ is represented in fig. 1.2-5.

The coordination of the higher BZ to the elementary region is not as easy as in the one-dimensional case represented in fig. 1.2-3. Because of the periodicity property (12) of the energy function, we may relate all regions outside the first BZ by a shift of vector $\boldsymbol{G}$ on it. The advantage of this procedure consists in the fact that the parts of the energy function from the different shifted regions may be combined to form an energy function continuous for symmetry

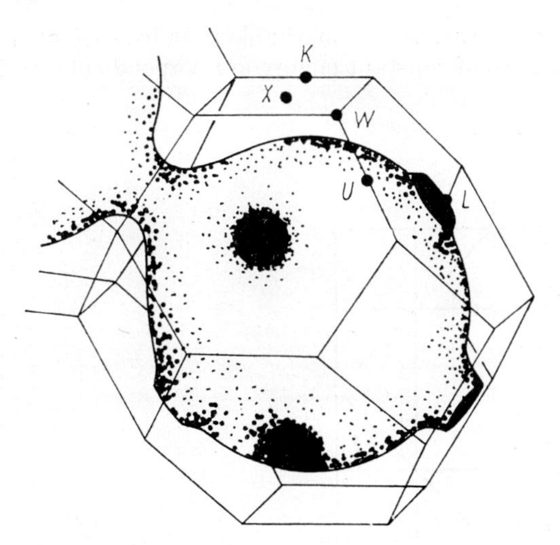

Fig. 1.2-5. First Brillouin zone and Fermi surface of copper. At the upper left, the neighbouring BZ of the periodic scheme is indicated. The marked notations of the high-symmetry points X, U, W, K, and L are used in fig. 1.2-7

reasons. This reduction on the first BZ allows transition to the function $\varepsilon_\nu(\boldsymbol{k})$ continuous in the first BZ but multivalued:

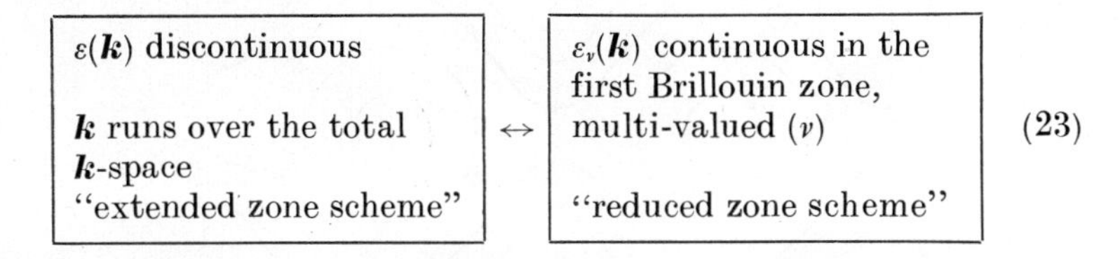

$\varepsilon(\boldsymbol{k})$ discontinuous $\boldsymbol{k}$ runs over the total $\boldsymbol{k}$-space "extended zone scheme"	$\leftrightarrow$	$\varepsilon_\nu(\boldsymbol{k})$ continuous in the first Brillouin zone, multi-valued (ν) "reduced zone scheme"	(23)

According to (12), the function $\varepsilon_\nu(\boldsymbol{k})$ in the reduced scheme is periodic. If this property is used, it is called a "periodic zone scheme".

The centre of the basic region (the first BZ) may be shifted to other points in $\boldsymbol{k}$-space. This shift is advantageous for states near the top of a band as is shown in fig. 1.2-6. To demonstrate this shifting, in fig. 1.2-6a) two-dimensional square lattice together with some low-indexed discontinuity lines is represented. The lattice points are indicated by small circles. The discontinuities are situated on the full lines. The inner square denotes the first BZ, the second consists of the four hatched regions which are shifted by the lattice vectors indicated by arrows into the basic region. They touch each other continuously at the broken lines. The eight parts of the third BZ are denoted by 3. In fig. 1.2-6b), lines of constant energy are drawn in the first BZ. Near the origin they are still spherical (as for free electrons), but near the zone boundaries strong deviations appear. It is remarkable that because of the property contained in (13), the lines of constant energy are perpendicular to the boundary of BZ.

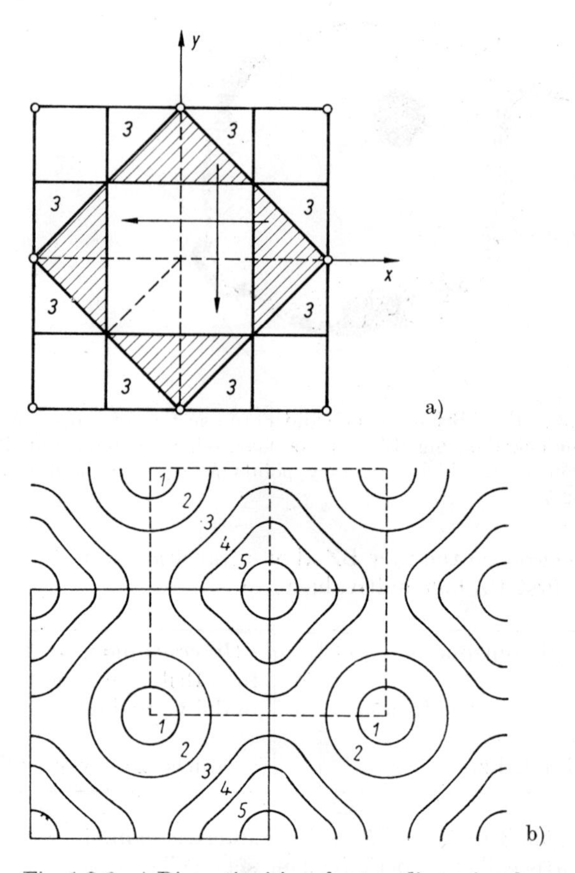

Fig. 1.2-6. a) Discontinuities of a two-dimensional square lattice
b) Lines of constant energy in the periodic scheme; they are numbered with increasing energy

The obvious representation of the three-dimensional dispersion relation is possible by means of surfaces of constant energy. The number $\mathrm{d}\mathcal{N}$ of states between two surfaces of constant energy with a difference $\mathrm{d}\varepsilon$ is denoted as density of states (DOS) $D(\varepsilon)$:

$$D(\varepsilon) = \frac{\mathrm{d}\mathcal{N}}{N\,\mathrm{d}\varepsilon}. \tag{24}$$

With the volume $\Delta^3 k$ from (10) which is available in $\boldsymbol{k}$-space for one state, we obtain in analogy to (1.1-4) for the number $\mathrm{d}\mathcal{N}$ of states in a volume element d^3k

$$\mathrm{d}\mathcal{N} = \frac{2\Omega}{(2\pi)^3}\,\mathrm{d}^3k\,. \tag{25}$$

The factor 2 arises as a consequence of the two possible spin orientations. Figures 1.2-7, 3.2-2b), 3.3-1, and 3.3-2b) contain examples for calculated densities of states.

A surface of constant energy with special importance is the Fermi surface given by

$$\varepsilon(\boldsymbol{k}) = \varepsilon_{\mathrm{F}}\,. \tag{26}$$

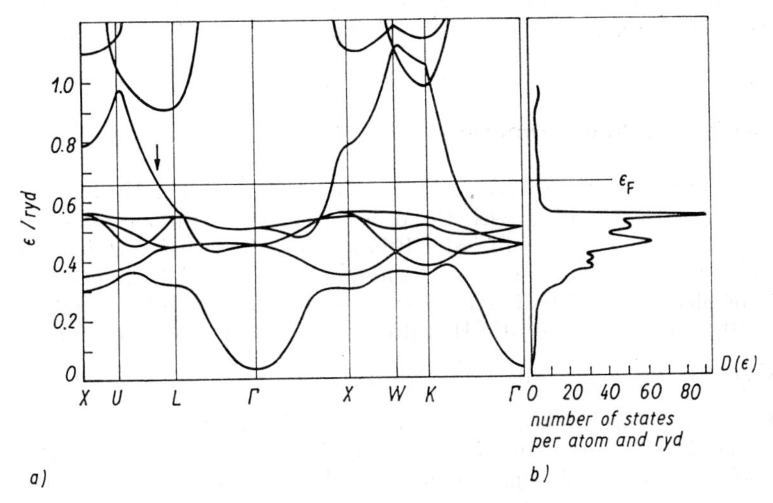

Fig. 1.2-7. a) Band structure of copper (after Woittennek 75). The point Γ denotes the centre of the BZ; the positions of the points X, W, K, U, and L on its boundary are explained in fig. 1.2-5.
b) Density of states $D(\varepsilon)$ of copper. The strong peaks are caused by the d-bands

As for the gas of free electrons in [1.1.2.], this separates regions which are empty or occupied with electrons in the ground state of the total system. Therefore, the Fermi energy will be determined by the condition that the available electrons just occupy all states of the volume included by the Fermi surface — the Fermi body. For free electrons, the Fermi surface is a spherical surface $\boldsymbol{k}^2 = k_{\mathrm{F}}^2$, whereas for lattice electrons (especially if several conduction electrons per atom are available) it deviates considerably from a sphere. As an example easy to study, the Fermi surface of copper (with one conduction electron per atom) is given in fig. 1.2-5. Experimental methods for determination of Fermi surfaces will be discussed in [4].

A further possibility for representation of the dispersion relation $\varepsilon_\nu(\boldsymbol{k})$ has been used in fig. 1.2-7. There, a one-dimensional plot of $\varepsilon_\nu(\boldsymbol{k})$ along lines between points of BZ with high symmetry is used.

1.3 Many-Particle Problems

The description of electronic structure of solids in a one-particle picture needs more thorough justification, as a solid is a macroscopic system of N atoms where N is a number of order 10^{20}. The simplifications arising from the regular structure of a crystal have already been discussed in the previous section. Here, we want essentially to investigate the separation of the motion of electrons and ions, and the mutual interaction of the electrons (for the latter see Ziesche, Lehmann and Eschrig 83, Ziesche and Paasch 83, Rennert and Paasch 83, for example).

1.3.1 Adiabatic Approximation

In general, a solid may be described by a non-relativistic theory as the particle energies responsible for binding with the order of 1 *ryd* are small compared with relativistic electron energies of order 10^4 *ryd* (mc^2). A more detailed treatment is necessary merely for spin-orbit interaction as a relativistic effect (important for electronic properties of heavier elements with order numbers higher than 40). The non-relativistic Hamiltonian H_{tot} of the total system

$$H_{\text{tot}} = T_{\text{n}} + T_{\text{e}} + V_{\text{nn}} + V_{\text{ne}} + V_{\text{ee}} \tag{1}$$

consists of the operators of kinetic energies of nuclei T_{n} and electrons T_{e} and of the energies of mutual interaction of nuclei and electrons: V_{nn}, V_{ne}, and V_{ee}. Under normal conditions some of the electrons will be bound more tightly to the nuclei as in free atoms, and as a result ions will be formed which are perturbed only very weakly by the motion of the residual electrons (the valence electrons).

The quantum mechanical problem of a system consisting of electrons and ions may be separated approximately, because of the large mass difference between them, into a problem of mutual interacting electrons in a static external field with Hamiltonian H',

$$H' = T_{\text{e}} + V_{\text{ee}}(\boldsymbol{r}_1, \boldsymbol{r}_2, \ldots) + V_{\text{ie}}(\boldsymbol{r}_1, \boldsymbol{r}_2, \ldots, \boldsymbol{R}_1, \boldsymbol{R}_2, \ldots), \tag{2}$$

and a problem of ions moving with an effective interaction $E_n(\boldsymbol{R}_1, \boldsymbol{R}_2, \ldots)$ depending on total electronic state n,

$$H_{\text{i}} = T_{\text{i}} + V_{\text{ii}}(\boldsymbol{R}_1, \boldsymbol{R}_2, \ldots) + E_n(\boldsymbol{R}_1, \boldsymbol{R}_2, \ldots). \tag{3}$$

The separation just described is called the adiabatic approximation. It was established for molecules by Born and Oppenheimer (27). For the wave function

$\Phi(\boldsymbol{r}_1, \boldsymbol{r}_2, \ldots, \boldsymbol{R}_1, \boldsymbol{R}_2, \ldots)$ of the total system (1), it means factorization

$$\Phi(\boldsymbol{r}_1, \boldsymbol{r}_2, \ldots, \boldsymbol{R}_1, \boldsymbol{R}_2, \ldots) = \Phi^{\mathrm{i}}(\boldsymbol{R}_1, \boldsymbol{R}_2, \ldots)\, \Psi_n^{\mathrm{e}}(\boldsymbol{r}_1, \boldsymbol{r}_2, \ldots, \boldsymbol{R}_1, \boldsymbol{R}_2, \ldots) \tag{4}$$

into an electronic part $\Psi_n^{\mathrm{e}}(\boldsymbol{r}_1, \boldsymbol{r}_2, \ldots, \boldsymbol{R}_1, \boldsymbol{R}_2, \ldots)$ corresponding to H' from (2), and an ionic part $\Phi^{\mathrm{i}}(\boldsymbol{R}_1, .\boldsymbol{R}_2, \ldots)$. The ionic coordinates $\boldsymbol{R}_1, \boldsymbol{R}_2, \ldots$ appear in H' and Ψ_n^{e} as external parameters. Therefore, the eigenvalues $E_n(\boldsymbol{R}_1, \boldsymbol{R}_2, \ldots)$ of the Schroedinger equation belonging to (2) also depend on $\boldsymbol{R}_1, \boldsymbol{R}_2, \ldots$ They act as an additional restoring potential for the ionic motion in (3). Putting in the equilibrium coordinates of the ion in the crystal from (2), the electronic band structure is finally obtained. For small deviations from the equilibrium positions, (3) yields the excitation spectrum of lattice vibrations — the phonons.

The physical basis of the adiabatic approximation consists in the following: because of the large mass difference between electrons and ions, the ionic motion measured on an electronic time scale is very slow, i.e. quasistatic. Therefore, the electronic system remains in the same quantum state Ψ_n^{e} (which itself depends parametrically on the ionic positions) during the motion of the ions.

The ratio $\alpha = m/M$ between electronic and ionic masses m and M, respectively, represents the small parameter of the adiabatic approximation (4). It follows from the equality of kinetic and potential energy of the ionic motion (harmonic) that the elongation of the ions from the equilibrium position is proportional to $\alpha^{1/4}$ and that the ratio of the energies of lattice vibrations $\hbar\omega_{\mathrm{ph}}$ and electronic energies (Fermi energy ε_{F}) is

$$\hbar\omega_{\mathrm{ph}}/\varepsilon_{\mathrm{F}} \sim \alpha^{1/2}. \tag{5}$$

The corrections on the adiabatic approximation induce the interaction of electrons and phonons.

A complete analysis of the problem for metals has been given by means of many-particle perturbation theory by Brovman and Kagan (74). They showed that the non-adiabatic corrections of the phonon frequencies are proportional to $\alpha^{1/4}$, and only in a small vicinity around the Migdal-Kohn anomalies (see [6]) are they of order $\alpha^{1/2}$. The finite phonon life times as a consequence of the electron-phonon interaction are proportional to $\alpha^{1/2}$. On the other hand, the corrections to the electronic energies near the Fermi energy (more exactly for $|\varepsilon(\boldsymbol{k}) - \varepsilon_{\mathrm{F}}| < \hbar\omega_{\mathrm{ph}}$) are large. Therefore, they must be taken into account more exactly in calculating measured quantities (as an example, see the electronic contribution to specific heat [5.1.2]).

1.3.2 Foundation of the One-Particle Picture

For further simplification of the many-particle problem (2) of interacting electrons in an external field, there are three options:

1. direct self-consistent approximations as they are known, such as the Hartree-Fock approximation from the physics of atoms and molecules;
2. density functional formalism;
3. the method of Green functions closely continuing the quasi-particle concept of the interacting electron gas in a constant external potential.

The first two approaches yield mostly approximate values for total energy and charge density, whereas the latter allows calculation of excitation energies.[1]) However, Hedin and Lundqvist (69) were able to show that at least for metals, each of them yields similar effective potentials[2]) justifying a self-consistent calculation in a one-particle picture.

Using the method of Green functions, a foundation of the Landau concept of quasi-particles (see Landau and Lifshitz 80) could also be given which on the other hand yields an explanation for the fact that in spite of strong mutual Coulomb interaction of electrons, they can be treated as nearly independent particles. The interaction solely results in a finite life time $\tau \sim (k - k_F)^{-2}$ of states with an energy $\varepsilon(\boldsymbol{k})$, and at the Fermi energy it becomes even infinite, since particles of that energy, as a consequence of the Pauli principle and the conservation of energy and momentum, cannot be scattered in practice by Coulomb interaction. The consequences of the resulting step in the momentum distribution function of interacting electrons at the Fermi momentum will be discussed in [4.2.1].

A recent progress within the electron theory of semiconductors is the solution of the mentioned gap problem. Using the Green's function method via an appropriate approximation for the exchange and correlation (XC) part of the self energy operator, Σ_{XC}, the energy gap between the valence and the conduction band can be calculated nowadays ab-initio in good agreement with the experiment (Godby et al. 87, 88, Hybertsen and Louie 87, Horsch et al. 87).

[1]) If the Hartree-Fock approximation is used to calculate also excitation energies for extended solids, then no metallic behaviour would be obtained theoretically: The DOS at the Fermi energy would vanish, i.e. no metals would exist theoretically.

If the density functional formalism is used to calculate also excitation energies for semiconductors, then the theoretical gap between the valence and the conduction band proves to be too small in comparison with the experimental value by a factor of about 1/2 ("gap problem").

[2]) Recent more refined or detailed considerations in this connection concern certain differences between the Fermi surfaces obtained from density functional formalism (Kohn-Sham calculations) and from experiment, the latter to be compared with results of quasi-particle calculations (Mearns 88).

In the Hartree-Fock approximation, the total wave function Ψ^e of the electronic system is written as a determinant of one-particle functions $\psi_n(\boldsymbol{r}_i)$. Therefore, Ψ^e obeys Pauli's principle. To the one-particle potential in (1.2-1) the remaining electrons contribute the Coulomb (or Hartree) potential

$$v_c(\boldsymbol{r}) = \int d^3r' n(\boldsymbol{r}') \frac{\epsilon^2}{|\boldsymbol{r} - \boldsymbol{r}'|} \tag{6}$$

with density $n(\boldsymbol{r})$ of electrons in the occupied one-particle states $\psi_n(\boldsymbol{r})$

$$n(\boldsymbol{r}) = \sum_n^{\text{occ}} \psi_n^*(\boldsymbol{r})\, \psi_n(\boldsymbol{r}) \tag{7}$$

and the exchange potential

$$\hat{v}_x \psi_n(\boldsymbol{r}) = -\epsilon^2 \int d^3r' \sum_{n'}^{\text{occ}} \frac{\psi_{n'}^*(\boldsymbol{r}')\, \psi_{n'}(\boldsymbol{r})}{|\boldsymbol{r} - \boldsymbol{r}'|}\, \psi_n(\boldsymbol{r}') \,. \tag{8}$$

Here n and $\boldsymbol{r}$ denote state and position variables of the electrons (both symbolically including spin states and variables). (8) represents an integral operator $\hat{v}_x$, not treated easily, which describes the tendency of electrons in the same spin state to avoid each other as a consequence of Pauli's principle. With it, the result also depends on the state ψ_n on which the operator acts (non-local exchange). Averaging (8) over all occupied states of the homogeneous electron gas yields the simple expression

$$v_x^S\big(n(\boldsymbol{r})\big) = -3 \left(\frac{3n(\boldsymbol{r})}{8\pi}\right)^{1/3} \epsilon^2 . \tag{9}$$

This was proposed by Slater (51) also for inhomogeneous systems (Slater exchange). Such a local exchange potential has proved to be successful in many calculations of total energies, ionization energies, and charge densities of atomic systems.

The density functional formalism is due to Hohenberg and Kohn (64). They proved that the wave function and thus each property of the ground state of a many-particle system in a given external potential $v_{ex}(\boldsymbol{r})$ are unique functionals of the particle density $n(\boldsymbol{r})$ alone and that the ground state energy may be written as

$$E_v[n] = \int d^3r v_{ex}(\boldsymbol{r})\, n(\boldsymbol{r}) + F[n] \tag{10}$$

where $F[n]$ represents a universal functional of the particle density $n(\boldsymbol{r})$, i.e. it is independent of the external potential $v_{ex}(\boldsymbol{r})$. However, F is unknown and contains the whole many-particle problem. The exact density $n(\boldsymbol{r})$ resulting in the external potential $v_{ex}(\boldsymbol{r})$ minimizes the functional (10) for fixed $v_{ex}(\boldsymbol{r})$ and for given particle number.

Kohn and Sham (65) were able to show that using the variational ansatz

$$n(\boldsymbol{r}) = \sum_n^{\text{occ}} |\psi_n(\boldsymbol{r})|^2 \tag{7'}$$

with test functions $\psi_n(\boldsymbol{r})$ to be varied, just the Schroedinger equation (1.2-1) results via a functional derivative of $F[n]$ and that the potential consists of an external part $v_{\mathrm{ex}}(\boldsymbol{r})$ (of the nuclei), the Coulomb part (6) $v_{\mathrm{c}}(\boldsymbol{r})$ of the electrons and the so-called exchange and correlation potential $v_{\mathrm{xc}}(\boldsymbol{r})$,

$$v(\boldsymbol{r}) = v_{\mathrm{ex}}(\boldsymbol{r}) + v_{\mathrm{c}}(\boldsymbol{r}) + v_{\mathrm{xc}}(\boldsymbol{r}) . \tag{11}$$

For the minimum of (10), one obtains as density $n(\boldsymbol{r})$ just the sum (7) over the occupied one-particle states.

For practical calculations, approximations for $F[n]$ have to be used. Kohn and Sham (and also Gaspar) used as the simplest approximation the dependency on the density known from the homogeneous electron gas (jellium):

$$v_{\mathrm{xc}}(\boldsymbol{r}) = \left.\frac{\left(\mathrm{d}n\varepsilon_{\mathrm{XC}}^{h}(n)\right)}{\mathrm{d}n}\right|_{n=n(\boldsymbol{r})},$$

where $\varepsilon_{\mathrm{XC}}^{h}(n)$ is the XC energy of the homogeneous jellium with density n. The best approximation for $\varepsilon_{\mathrm{XC}}^{h}(n)$ follows from the quantum Monte Carlo calculations of Ceperley and Alder by means of an appropriate interpolation or parametrization (Vosko et al. 80, Perdew and Zunger 81, Aguilera-Navarro et al. 85). In the Hartree-Fock approximation (i.e. only considering the exchange energy), the expression

$$v_{\mathrm{xc}}(\boldsymbol{r}) \approx v_{\mathrm{X}}^{\mathrm{KSG}}(\boldsymbol{r}) = \frac{2}{3}\, v_{\mathrm{X}}^{\mathrm{S}}\big(n(\boldsymbol{r})\big) \tag{12}$$

with $v_{\mathrm{X}}^{\mathrm{S}}$ from (9) is obtained.

In the theory of the electron gas, the total energy may be expressed by the dielectric function ε from [5.2.1]. By means of the best approximations known so far, Hedin and Lundqvist (71) obtained the so-called local density approximation (LDA)[1])

$$v_{\mathrm{xc}}(\boldsymbol{r}) = \alpha(r_s)\, v_{\mathrm{X}}^{\mathrm{S}}(n), \tag{13}$$

$$\alpha(r_s) = \frac{2}{3} + 0.0246 r_s \ln\left(1 + 21/r_s\right).$$

For metallic densities, the factor α is nearly equal to 0.8. Each of the expressions (9), (12), and (13) has the form $\alpha v_{\mathrm{X}}^{\mathrm{S}}$ with values of α between 2/3 and 1. This was the reason for introducing the X_α method (see Slater 74), in which α represents a parameter to be fitted to the results of Hartree-Fock calculations for isolated atoms. Its values lie (excluding hydrogen) in the region between 0.7 and 0.8. Also (13) and the method of Green functions yield results almost in agreement, but they were obtained in very different manners. The results show that for the calculation of the total energy and density and for the

[1]) Recent surveys on LDA are given by Lundqvist and March (83) and Dahl and Avery (84).

calculation of the excitation (quasi-particle) spectrum of metals, the same effective exchange and correlation potential may be used, and that the X_α expression contains much more correlation than the Hartree-Fock approximation. On this fact, the reason for the large practical success of the X_α method is based.

The basic equations for a self-consistent calculation of the crystal potential are (7), (11), and (13) in connection with the Schroedinger equation (1.2-1). At first, the last equation is solved with a starting potential resulting from an overlap of the atomic contributions. From that, ψ_n and $n(\boldsymbol{r})$ and via (11) the improved potential follow. This cycle is repeated several times until self-consistency is reached. In this way, the crystal potential for a large number of pure metals has been calculated (see for example Moruzzi et al. (78)).

In analogy to (13), Barth and Hedin (72) calculated formulae for spin-polarized systems. Then α^+ and α^- and their dependence on n^+ and n^- according to the orientation of the spin have to be taken into account. This opens up the understanding and calculation of cooperative magnetic phenomena (such as ferromagnetism) from first principles. The spatial behaviour of magnetic moments may be calculated via $(n^+ - n^-)$. This quantity can be measured by means of elastic scattering of polarized neutrons (magnetic form factor). The local magnetic moments per atom are obtained by an integration of $(n^+ - n^-)$ over the WS cell. The results gained are presented in brief in [3.2].

The eigenvalues prove to be less sensitive to approximations for the potential than the eigenfunctions. Phenomena which require the wave functions for a theoretical description are the key to determine the validity limits of the one-particle model, and a deeper understanding of many-particle effects. Unfortunately, so far only a few experiments have revealed detailed information on wave functions in metals. The form factors measured by neutron diffraction could be an exception to this situation.

LDA and its improvements correspond to the case of weak electron correlation. But there are systems of recent interest characterized by strong electron correlation. Examples are high T_c superconductors, heavy Fermion systems, mixed valence compounds. They seem to need an ab-initio theory of strong electron correlation for a full understanding. Up till now only models exist for that case, e.g. the Hubbard model or the t-J-model.

2 Methods of Band Structure Calculations

The band structure contains essential information on the electronic structure. This statement holds especially well for systems and properties where the excitation spectrum can be described in the one-particle approximation. Therefore, we assume in this chapter the one-particle potential $v(\boldsymbol{r})$ to be given including many-particle effects [1.3.2]. Then the solution of the Schroedinger equation (1.2-1) yields eigenvalues $\varepsilon_\nu(\boldsymbol{k})$ and wave functions $\psi_{\boldsymbol{k}\nu}(\boldsymbol{r})$ which on the other hand allow us to calculate the properties of the system.

The periodicity (1.2-3) of the potential presents the essential advantage that the problem has to be solved only in the elementary cell and that this solution may be extended via the Bloch theorem (1.2-5) into all other regions of the crystal.

The single developed methods of band structure calculation correspond to different physical situations in the solid:

— the tight-binding (TB) method starts from an ensemble of weakly interacting atoms,
— the pseudopotential method starts from nearly free electrons (NFE).

More rigorous methods including both limiting cases are

— the APW method and
— the KKR method.

An approach describing both tightly bound as well as nearly free electrons including their hybridization (H) in a transparent physical manner may be derived from the KKR method, and is called

— the H-NFE-TB method.

For alloy systems with crystalline disorder,

— the coherent potential approximation (CPA) has been developed.

Essentials of the first four approaches have already been worked out in the fourties and fifties. However, they first became important 20 years ago, when experimental results on the band structure and especially on Fermi surface and with that, reliable information on the crystal potential, became available. A tight interplay between theory and experiment allowed us to understand

and prove why conduction electrons in metals behave like nearly free electrons. With the increasing capacity of computers at the end of the sixties the APW and KKR methods which demand considerable numerical expense have been widely used. The desire for decreasing costs and treating more complicated systems quickly and with a greater variety led in the middle of the seventies to the development of so-called linear methods (LAPW, LKKR, LMTO). In this text we treat as a typical representative, the H-NFE-TB method. Relativistic versions of nearly all these methods have been developed to include the spin-orbit interaction.

2.1 Tight-Binding Method

2.1.1 Atomic Wave Functions and Their Linear Combinations

The method starts with the idea that the neighbours of a given atom represent only a weak perturbation of this atom. This picture can be described most easily by means of the Wannier representation (1.2-5″) in the case where the Wannier function $w(\boldsymbol{r} - \boldsymbol{R})$ is very well localized at the lattice site $\boldsymbol{R}$ and its overlap with neighbouring wave functions is small. In the simplest, zeroth approximation, $w(\boldsymbol{r})$ will be represented by the atomic wave function $\varphi_n^{\text{at}}(\boldsymbol{r})$ of the n-th level:

$$w_n(\boldsymbol{r}) \approx \varphi_n^{\text{at}}(\boldsymbol{r}) . \tag{1}$$

This approximation breaks down if several atomic levels have nearly the same energy or are immediately degenerated. Then a linear combination of these atomic orbitals has to be used:

$$\sum_{n'} \varphi_{n'}^{\text{at}}(\boldsymbol{r}) \, c_{n'n} . \tag{2}$$

This procedure is thus called the *LCAO method*. In the sum over n', all atomic orbitals in the considered energy region have to be included. Treating d-electrons, this means at least five states.

The LCAO method represents one of the oldest procedures for calculating the band structure (Slater and Koster 54). For a detailed presentation, we use the modern version of Eschrig (83a, 88), which overcomes a series of disadvantages of the original method.

The wave functions $\varphi_n^{\text{at}}(\boldsymbol{r})$ are eigenfunction of an "atomic" Hamiltonian $H^{\text{at}} = t + v^{\text{at}}(\boldsymbol{r})$,

$$\big(t + v^{\text{at}}(\boldsymbol{r})\big) \, \varphi_n^{\text{at}}(\boldsymbol{r}) = \varphi_n^{\text{at}}(\boldsymbol{r}) \, \varepsilon_n , \tag{3}$$

with eigenvalues ε_n. In the vicinity of a given ion at site $\boldsymbol{R} = 0$, the potential should approach as closely as possible the crystal potential. Outside the atomic region, it can be chosen almost arbitrarily, however. The functions φ_n^{at} at a given site are of course orthogonal,

$$\int \mathrm{d}^3 r \varphi_n^{\mathrm{at}*}(\boldsymbol{r})\, \varphi_{n'}^{\mathrm{at}}(\boldsymbol{r}) = \delta_{nn'}, \tag{4}$$

and complete. We classify them in strongly localized core states (index c) and more extended valence states (index n). Therefore, the overlap of the core states at neighbouring lattice sites can be neglected in the following:

$$\int \mathrm{d}^3 r \varphi_c^{\mathrm{at}*}(\boldsymbol{r})\, \varphi_c^{\mathrm{at}}(\boldsymbol{r} - \boldsymbol{R}) = \delta_{\boldsymbol{R},0}. \tag{5}$$

This is not valid for valence states. It is appropriate to orthogonalize them to the core states $\varphi_c^{\mathrm{at}}(\boldsymbol{r} - \boldsymbol{R})$ of neighbouring atoms

$$\varphi_n(\boldsymbol{r}) = \varphi_n^{\mathrm{at}}(\boldsymbol{r}) - \sum_{c,\boldsymbol{R}\neq 0} \varphi_c^{\mathrm{at}}(\boldsymbol{r} - \boldsymbol{R})\, a_{c,\boldsymbol{R}}^n. \tag{6}$$

Choosing coefficients $a_{c,\boldsymbol{R}}^n$ as

$$a_{c,\boldsymbol{R}}^n = \int \mathrm{d}^3 r \varphi_c^{\mathrm{at}*}(\boldsymbol{r} - \boldsymbol{R})\, \varphi_n^{\mathrm{at}}(\boldsymbol{r}) \tag{7}$$

guarantees the desired orthogonality for all $\boldsymbol{R}$

$$\int \mathrm{d}^3 r \varphi_n^{\mathrm{at}*}(\boldsymbol{r} - \boldsymbol{R})\, \varphi_n(\boldsymbol{r}) = 0 \quad \text{for all } \boldsymbol{R}. \tag{8}$$

The sums over the neighbouring sites $\boldsymbol{R}$ in (6) represent small corrections at the wave function which because of the strong potential in the immediate neighbourhood of the ions and because of the large binding energies ε_c of the core states may induce essential corrections of the energy. However, for increasing $|\boldsymbol{R}|$ the coefficients $a_{c,\boldsymbol{R}}^n$ decrease strongly. Using the $\varphi_n(\boldsymbol{r})$ from (6) at the different lattice sites in analogy to (1.2-5″), functions with the desired lattice symmetry may be constructed (Bloch sum)

$$\varphi_{n\boldsymbol{k}}(\boldsymbol{r}) = N^{-1/2} \sum_{\boldsymbol{R}} \mathrm{e}^{i\boldsymbol{k}\boldsymbol{R}} \varphi_n(\boldsymbol{r} - \boldsymbol{R}). \tag{9}$$

Corresponding to (2) a superposition of the "atomic orbitals" contained in (9) yields the wave functions of the crystal

$$\psi_{\boldsymbol{k}\nu}(\boldsymbol{r}) = \sum_{n'} c_{\nu n'}(\boldsymbol{k})\, \varphi_{n'\boldsymbol{k}}(\boldsymbol{r}). \tag{10}$$

The summation in (10) runs over all states n' whose energy is located in the considered energy region. The coefficients $c_{\nu n'}(\boldsymbol{k})$ have to be determined via the variational principle (11). The representation (10) of the Bloch states is exact since the functions φ_n^{at} form a complete system[1]). Because of the previous orthogonalization (6) to the core states, they may be excluded in (10). Since in consequence the sum over n' in (10) includes only a few valence states in the interesting energy region in versions (6), (9) and (10), the numerical expense for high numerical accuracy may be reduced considerably.

[1]) In practice, the wave function (10) is overcomplete, as the $\varphi_n^{\mathrm{at}}(\boldsymbol{r} - \boldsymbol{R})$ form a complete system for each single $\boldsymbol{R}$. Limiting to a finite set of n, this question is only academic.

The minimization of the energy ε by variation of the $c_{\nu n'}(\boldsymbol{k})$

$$\delta\varepsilon = 0 \quad \text{with} \quad \varepsilon = \int \mathrm{d}^3 r \psi^*_{\boldsymbol{k}\nu}(\boldsymbol{r})\, H\psi_{\boldsymbol{k}\nu}(\boldsymbol{r}) \Big/ \int \mathrm{d}^3 r \psi^*_{\boldsymbol{k}\nu}(\boldsymbol{r})\, \psi_{\boldsymbol{k}\nu}(\boldsymbol{r}) \tag{11}$$

yields for the coefficients $c_{\nu n'}(\boldsymbol{k})$, the linear system of *LCAO equations*

$$\sum_{n'} \big(H_{n''n'}(\boldsymbol{k}) - \varepsilon_\nu(\boldsymbol{k})\, N_{n''n'}(\boldsymbol{k})\big)\, c_{\nu n'}(\boldsymbol{k}) = 0 \tag{12}$$

with

$$H_{n''n'}(\boldsymbol{k}) = \sum_{\boldsymbol{R}} \mathrm{e}^{i\boldsymbol{k}\boldsymbol{R}} \int \mathrm{d}^3 r \varphi^*_{n''}(\boldsymbol{r} + \boldsymbol{R})\, H\varphi_{n'}(\boldsymbol{r}) \tag{13$'$}$$

and

$$N_{n''n'}(\boldsymbol{k}) = \sum_{\boldsymbol{R}} \mathrm{e}^{i\boldsymbol{k}\boldsymbol{R}} \int \mathrm{d}^3 r \varphi^*_{n''}(\boldsymbol{r} + \boldsymbol{R})\, \varphi_{n'}(\boldsymbol{r}) \tag{13$''$}$$

and H from (15).

Using (6) and taking into consideration the strong localization of the core functions given by $H\varphi_c(\boldsymbol{r}) = \varepsilon_c\varphi_c(\boldsymbol{r})$, we obtain for the integrals in (13′) and (13″) the expressions

$$\begin{aligned} H^{\boldsymbol{R}}_{n''n'} &= \int \mathrm{d}^3 r \varphi^*_{n''}(\boldsymbol{r} + \boldsymbol{R})\, H\varphi_{n'}(\boldsymbol{r}) \\ &= \int \mathrm{d}^3 r \varphi^{\mathrm{at}*}_{n''}(\boldsymbol{r} + \boldsymbol{R})\, H\varphi_{n'}(\boldsymbol{r}) - \sum_{c,\boldsymbol{R}'}{}' \varepsilon_c a^{n''*}_{c\boldsymbol{R}} a^{n'}_{c(\boldsymbol{R}'-\boldsymbol{R})}, \end{aligned} \tag{14$'$}$$

$$\begin{aligned} N^{\boldsymbol{R}}_{n''n'} &= \int \mathrm{d}^3 r \varphi^*_{n''}(\boldsymbol{r} + \boldsymbol{R})\, \varphi_{n'}(\boldsymbol{r}) \\ &= \int \mathrm{d}^3 r \varphi^{\mathrm{at}*}_{n''}(\boldsymbol{r} + \boldsymbol{R})\, \varphi_{n'}(\boldsymbol{r}) - \sum_{c,\boldsymbol{R}'}{}' a^{n''*}_{c,\boldsymbol{R}'} a^{n'}_{c(\boldsymbol{R}'-\boldsymbol{R})}. \end{aligned} \tag{14$''$}$$

The prime on the sums in (14) means the term with $\boldsymbol{R}' = 0$ may be excluded because the integrals $a^n_{c,\boldsymbol{R}'}$ vanish for $\boldsymbol{R}' = 0$.

2.1.2 Appropriate Choice of Atomic Potential

Evaluation of the integrals $H^{\boldsymbol{R}}_{nn'}$ and $N^{\boldsymbol{R}}_{nn'}$ in (14) demands the highest numerical expense. For well localized atomic functions φ^{at}_n, we can restrict ourselves to small values of $\boldsymbol{R}$. Here, the option of the atomic potential (3) first introduced by Eschrig (82) gives a decisive advantage. The crystal potential $v(\boldsymbol{r})$ in the Hamiltonian H,

$$H = t + v(\boldsymbol{r}), \tag{15}$$

may be represented as an overlap of ionic potentials $\tilde{v}$ at the lattice sites $\boldsymbol{R}$:

$$v(\boldsymbol{r}) = \sum_{\boldsymbol{R}} \tilde{v}(|\boldsymbol{r} - \boldsymbol{R}|). \tag{16}$$

The use of $\tilde{v}$ in (3) leads to fundamental difficulties which were the reason for doubts in the validity of the approach. The reason lies in the inconvenient asymptotic behaviour of the ionic potential for large distances

$$\tilde{v} \sim r^{-1} \quad \text{for} \quad r \to \infty. \tag{17}$$

On the one hand this is the reason why the corresponding wave functions φ^{at} are not well localized and why, therefore, the sums in (13) converge very weakly. On the other hand, the states in the potential $\tilde{v}$ decompose into bound and free states. Even although because of (17) an infinite number of bound states exists, they do not form a complete system. Therefore, omitting the free states leads to fundamental difficulties since the wave functions of every bound state with an energy above the conduction band have nearly the same behaviour within the considered Wigner-Seitz cell. Only the free states have the required behaviour (with more nodes) in this region. Therefore, Eschrig chooses for the atomic potential

$$v^{\text{at}}(\boldsymbol{r}) = v(\boldsymbol{r}) + (r/r_0)^n . \tag{18}$$

Through this, all the mentioned disadvantages disappear. For $r \approx R_{\text{WS}}$ (radius of the Wigner-Seitz sphere), the solutions of (3) inside the atoms show the necessary behaviour of atomic wave functions, whilst outside they vanish sufficiently rapidly. In the potential (18) only bound states exist. The option $n = 4$ and $r_0 \approx R_{\text{WS}}$ proved to be advantageous for good convergency and for small sets of functions which covers the essential features for the wave functions of conduction bands. In this case, the sums over $\boldsymbol{R}$ in (13) have to include only first and second nearest neighbours. Three-centre integrals in (14′) of the type

$$\int \mathrm{d}^3r \varphi_n^{\text{at}*}(\boldsymbol{r} + \boldsymbol{R})\, v^{\text{at}}(\boldsymbol{r} + \boldsymbol{R}')\varphi_{n'}(\boldsymbol{r}) ,$$

usually demanding considerable numerical effort, may be neglected as a rule. In the atomic potential (18) the atomic wave functions $\varphi_n(\boldsymbol{r})$ in (6) are well localized. Well localized Wannier functions obeying the orthogonality relations (1.2-2″) may be generated by a linear combination of the $\varphi_n(\boldsymbol{r} - \boldsymbol{R})$:

$$w_\nu(\boldsymbol{r}) = \sum_{n',\boldsymbol{R}} \alpha_{\nu n'}^{\boldsymbol{R}} \varphi_{n'}(\boldsymbol{r} - \boldsymbol{R}) . \tag{19}$$

The coefficients $\alpha_{\nu n'}^{\boldsymbol{R}}$ die away very quickly.

2.1.3 Tight-Binding Approximation

For atomic states overlapping only weakly (such as the d-states) and, therefore, generating narrow bands only, eqn. (12) may be simplified further. In this case, all overlap integrals of neighbouring wave functions are neglected in (13″),

$$N_{n''n'}(\boldsymbol{k}) = \delta_{n''n'} . \tag{20}$$

Using the Schroedinger equations (3) and (14), (15), and (16), we obtain from (12) the so-called *tight-binding approximation* (TB)

$$\sum_{n'} \left[V_{n''n'}(\boldsymbol{k}) - \left(\varepsilon_\nu(\boldsymbol{k}) - \varepsilon_{n'}\right) \delta_{n''n'} \right] c_{\nu n'}(\boldsymbol{k}) = 0 , \tag{21}$$

$$V_{n''n'}(\boldsymbol{k}) = \sum_{\boldsymbol{R}} \mathrm{e}^{i\boldsymbol{k}\boldsymbol{R}} \int \mathrm{d}^3r \varphi_{n''}^{\text{at}*}(\boldsymbol{r} + \boldsymbol{R}) \left(\tilde{v}\boldsymbol{r}) - \tilde{v}(\boldsymbol{r})\right) \varphi_{n'}^{\text{at}}(\boldsymbol{r}) . \tag{22}$$

The summation n' includes in this case only the group of physically interesting bands. For relatively narrow valence bands in insulators well separated from the conduction band, the simple equations (21) usually represent a sufficient approximation. The d-bands in transition metals are well described in those cases where the respective small $\boldsymbol{k}$-region where the d-band hybridizes with the sp-band is unimportant.

Frequently, the TB approximation is used as an empirical interpolation scheme. Then the overlap integrals in (22) are considered to be fitting parameters. In the case of an isolated tightly bound band, (21) reduces to a single equation

$$\varepsilon_\nu(\boldsymbol{k}) = \varepsilon_\nu + \sum_{\boldsymbol{R}} \mathrm{e}^{i\boldsymbol{k}\boldsymbol{R}} V_\nu(\boldsymbol{R}), \tag{23}$$

$$V_\nu(\boldsymbol{R}) = \int \mathrm{d}^3 r \varphi_\nu^{\mathrm{at}*}(\boldsymbol{r} + \boldsymbol{R}) \left(v(\boldsymbol{r}) - \tilde{v}(\boldsymbol{r})\right) \varphi_\nu^{\mathrm{at}}(\boldsymbol{r}). \tag{23'}$$

Then the sum over $\boldsymbol{R}$ is restricted on the next nearest neighbours. ε_ν equals the energy of the atomic states forming the band.

In the cubic lattice, we have from symmetry reasons only two different overlapping integrals $V_\nu(0)$ and $V_\nu(\boldsymbol{R})$. $V_\nu(0)$ represents a shift of the atomic levels ε_ν induced by the crystal potential. $V_\nu(\boldsymbol{R})$ gives the band structure as a consequence of the overlap of wave functions and potential at neighbouring sites.

The LCAO method is well suited for the representation of local quantities; its advantages have been used mainly in theoretical chemistry for the understanding of covalent bonds (even in semiconductors). Therefore, many empirical or semi-empirical versions of LCAO exist.

In solid state physics, the LCAO method takes an increasing meaning for the description of phenomena in real space, such as in systems without long range order or for a more accurate consideration of the electron-electron interaction (e.g. Hubbard model).

2.2 Pseudopotential Method

2.2.1 Orthogonalized Plane Waves

We now consider the evaluation of the weak potential experienced by valence electrons which must be known in the nearly free electron (NFE) approximation. The basic idea follows from the fact that for many metals the wave function consists of two parts. In very large spatial regions, it behaves like a free plane wave (there the factor $u_{\boldsymbol{k}\nu}(\boldsymbol{r})$ in (1.2-5') is nearly constant, and this is shown for the wave function of metallic sodium in fig. 2.2-1). Only in the immediate surroundings of the ionic cores does it show an oscillatory form. In analogy to (2.1-6), this behaviour may be represented by the ansatz

$$\chi_{\boldsymbol{k}}(\boldsymbol{r}) = \mathrm{e}^{i\boldsymbol{k}\boldsymbol{r}} - \sum_{c} \varphi_{c\boldsymbol{k}}(\boldsymbol{r})\, a_c(\boldsymbol{k}). \tag{1}$$

The first term represents the plane wave whereas the second is given by a linear combination of the TB functions (2.1-9) of all bands in the crystal with energies lower than the considered one. Their atomic "core" functions $\varphi_{c\boldsymbol{k}}$ show well the desired oscillating behaviour in the vicinity of the ions. The fact that the functions $\chi_{\boldsymbol{k}}$ in this region exhibit the correct behaviour (this means the right nodal numbers corresponding to the valence state) is guaranteed by choosing the coefficients $a_c(\boldsymbol{k})$ so that the Pauli principle is fulfilled, i.e. the valence states $\chi_{\boldsymbol{k}}$ are orthogonal to the lower lying states $\varphi_{c\boldsymbol{k}}$:

$$a_c(\boldsymbol{k}) = \int \mathrm{d}^3 r \varphi^*_{c\boldsymbol{k}}(\boldsymbol{k})\, \mathrm{e}^{i\boldsymbol{k}\boldsymbol{r}}. \tag{2}$$

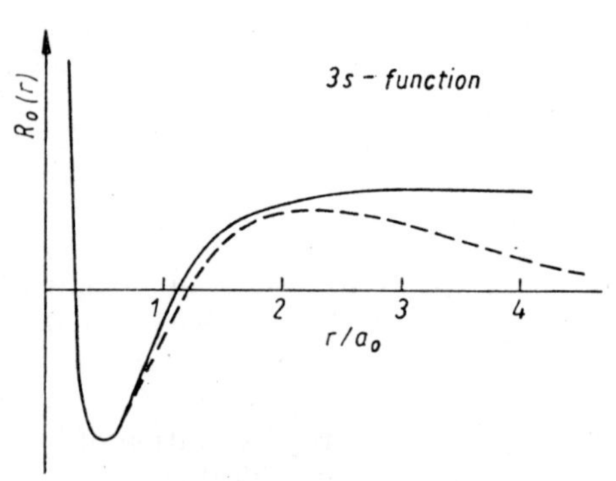

Fig. 2.2-1. Behaviour of the radial wave function of the valence state in sodium; in the crystal for $\boldsymbol{k} = 0$ (drawn up to the Wigner-Seitz radius) and in the free atom (broken line)

Therefore, the wave function $\chi_{\boldsymbol{k}}(\boldsymbol{r})$ defined by (1) is called the orthogonalized plane wave (OPW):

$$\chi_{\boldsymbol{k}}(\boldsymbol{r}) = \mathrm{e}^{i\boldsymbol{k}\boldsymbol{r}} - \sum_c \varphi_{c\boldsymbol{k}}(\boldsymbol{r}) \int \mathrm{d}^3 r' \varphi^*_{c\boldsymbol{k}}(\boldsymbol{r}')\, \mathrm{e}^{i\boldsymbol{k}\boldsymbol{r}'}. \tag{3}$$

The wave function of a valence electron is superposed as a linear combination of OPW's for the same eigenvalue $\boldsymbol{k}$ of the translation operator:

$$\psi_{\boldsymbol{k}}(\boldsymbol{r}) = \sum_{\boldsymbol{G}} \alpha_{\boldsymbol{k}+\boldsymbol{G}} \chi_{\boldsymbol{k}+\boldsymbol{G}}(\boldsymbol{r}). \tag{4}$$

The coefficients $\alpha_{\boldsymbol{k}+\boldsymbol{G}}$ and the energy eigenvalues $\varepsilon(\boldsymbol{k})$ are determined by minimizing the expectation value H in the state (4) via variation of the coefficients $\alpha_{\boldsymbol{k}+\boldsymbol{G}}$. We assume that the series (4) converges very fast so that the first term $\chi_{\boldsymbol{k}}$ describes the behaviour of the total function quite well in the lowest approximation and the few important residual coefficients $\alpha_{\boldsymbol{k}+\boldsymbol{G}}$ in (4) may be determined easily by perturbation theory. Thus the free electron approximation has been repeated on a new level of interpretation. The approach described is called the *OPW method*, and its first approximation where only one term $\chi_{\boldsymbol{k}}$ in (4) is used is called the one-OPW approximation.

2.2.2 Pseudopotentials

Let us now investigate the stated fast convergency of the sum (4). For this, we represent (4) using (3) as

$$\psi_{\boldsymbol{k}}(\boldsymbol{r}) = \Phi_{\boldsymbol{k}}(\boldsymbol{r}) + \varphi_{\boldsymbol{k}}(\boldsymbol{r}) \quad \text{with} \quad \Phi_{\boldsymbol{k}}(\boldsymbol{r}) = \sum_{\boldsymbol{G}} \alpha_{\boldsymbol{k}+\boldsymbol{G}}\, \mathrm{e}^{i(\boldsymbol{k}+\boldsymbol{G})\boldsymbol{r}}, \tag{5}$$

$$\varphi_{\boldsymbol{k}}(\boldsymbol{r}) = -\sum_{\boldsymbol{G},c} \varphi_{c\boldsymbol{k}}(\boldsymbol{r}) \int \mathrm{d}^3r' \varphi^*_{c\boldsymbol{k}}(\boldsymbol{r}')\, \mathrm{e}^{i(\boldsymbol{k}+\boldsymbol{G})\boldsymbol{r}'} \alpha_{\boldsymbol{k}+\boldsymbol{G}} = -\sum_c P_c \Phi_{\boldsymbol{k}} \tag{6}$$

with

$$P_c \Phi_{\boldsymbol{k}}(\boldsymbol{r}) = \varphi_{c\boldsymbol{k}}(\boldsymbol{r}) \int \mathrm{d}^3r' \varphi^*_{c\boldsymbol{k}}(\boldsymbol{r}')\, \Phi_{\boldsymbol{k}}(\boldsymbol{r}') \tag{6'}$$

whereas $\Phi_{\boldsymbol{k}}(\boldsymbol{r})$ contains the smooth part of the wave functions. P_c denotes the operator projecting out from the function, $\Phi_{\boldsymbol{k}}(\boldsymbol{r})$ the contributions of the core functions $\varphi_{c\boldsymbol{k}}(\boldsymbol{r})$. Writing, instead of the Schroedinger equation (1.2-1) for $\psi_{\boldsymbol{k}}$, an equivalent differential equation for the pseudo wave function $\Phi_{\boldsymbol{k}}$ using (2), (5) and (6),

$$h\Phi_{\boldsymbol{k}} = \Phi_{\boldsymbol{k}}\varepsilon(\boldsymbol{k}), \quad h = -\frac{\hbar^2}{2m}\frac{\partial^2}{\partial \boldsymbol{r}^2} + w(\boldsymbol{r}), \tag{7}$$

one recognizes clearly that $\Phi_{\boldsymbol{k}}$ is a smooth function, i.e. the pseudopotential $w(\boldsymbol{r})$ really represents only a weak perturbation of the Hamiltonian $-\frac{\hbar^2}{2m}\frac{\partial^2}{\partial \boldsymbol{r}^2}$ for free electrons. In connection with (8) and (9), equation (7) is to be referred to as the Phillips-Kleinman equation. The transformation of (1.2-1) into (7) yields for the pseudopotential

$$w(\boldsymbol{r}) = v(\boldsymbol{r}) + \hat{v}_R(\boldsymbol{r}) \tag{8}$$

with

$$\hat{v}_R(\boldsymbol{r}) = \sum_c \left(\varepsilon(\boldsymbol{k}) - \varepsilon_c\right) P_c\,. \tag{9}$$

According to (8), $w(\boldsymbol{r})$ consists of the ionic potential $v(\boldsymbol{r})$ and the so-called repulsion potential $\hat{v}_R(\boldsymbol{r})$ compensating each other approximately. Correspondingly to (9), $\hat{v}_R(\boldsymbol{r})$ represents a non-local potential containing P_c as an integral operator acting on the pseudo wave function $\Phi_{\boldsymbol{k}}$. Furthermore, it depends on the energy eigenvalue $\varepsilon(\boldsymbol{k})$ still to be calculated and on the known core energies ε_c. $\hat{v}_R$ guarantees that because of the Pauli principle, the valence electrons penetrate only very weakly into the core region and, therefore, are unable to feel the strong ionic potential $v(\boldsymbol{r})$.

The sketched approach is called *pseudopotential method* (Harrison 66). It is well suited for band structure calculation in so-called simple metals and semiconductors such as silicon. It yields good agreement with band structures determined experimentally. In simple metals the conduction and valence bands arise from atomic s- and p-states.

Applying the pseudopotential method, calculation of band structure from first principles is possible (for the consequent application of the approach on pure metals and alloys see Hafner 77, 79, and 87, for example). On the other hand, in the method of model potentials described in the following chapter, certain information (e.g. concerning hardly calculable many body effects) on atomic potentials from experimental data obtained on free atoms and ions, respectively, is used. With it, the numerical effort reduces considerably.

2.2.3 Model Potentials

In the method of model potentials (Heine and Abarenkov 64) related to the pseudopotential method, a potential in the shape (fig. 2.2-2)

$$w^0(r) = -Z\,\epsilon^2 \begin{cases} 1/r \\ A_l(\varepsilon) \end{cases} \quad \text{for} \quad r \gtrless R_l \tag{10}$$

is used. In the external region $r > R_l$, it has the same behaviour as a Z-fold charged ion. In the inner region the potential corresponding to the non-local expression (9) is represented by constants $A_l(\varepsilon)$ depending on angular momentum l and energy ε of the state to be calculated. Therefore, the model potential is also non-local, and means that it acts on different states differently. $A_l(\varepsilon)$ may be determined, for example, from spark spectra of free ions covering the interesting energy region. In this procedure, use is made of the prescription that the potential (10) has to reproduce the experimental one-electron spectra of the ion. In the external region $r > R_l$, the pseudo wave function belonging to the potential (10) behaves as the true wave function. However, contrary to the true wave function the pseudo wave function has no nodes in the inner region. Therefore, the model potential turns out essentially smoother than the true potential.

In a crystal, the resulting potential becomes still smoother because of the overlapping contributions of neighbouring atoms (fig. 2.2-2) and because of

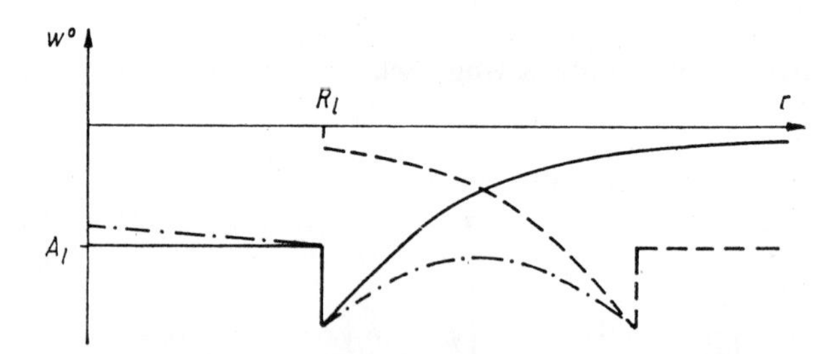

Fig. 2.2-2. Model potential $w^0(r)$ of Heine and Abarenkov according to (2.2-10) for an ion at $r = 0$ (straight line) and a neighbouring ion (broken line) and their superposition (dash-dotted line, excluding a constant contribution)

the screening of the long range Coulombic part $\sim r^{-1}$ of the ions by the gas of the nearly free conduction electrons.

The Fourier transform of the crystal potential

$$W(\boldsymbol{r}) = \sum_{\boldsymbol{R}} w(\boldsymbol{r} - \boldsymbol{R}) \tag{11}$$

follows in analogy to (1.2-21) as

$$W(\boldsymbol{k}) = \frac{1}{\Omega} \int d^3r \, e^{-i\boldsymbol{kr}} W(\boldsymbol{r}) = w(\boldsymbol{k}) \, S(\boldsymbol{k}) \tag{12}$$

with $w(\boldsymbol{k})$ as the Fourier transform of a single screened ion potential $w(\boldsymbol{r})$, the *atomic form factor*

$$w(\boldsymbol{k}) = \frac{1}{\Omega_0} \int d^3r \, e^{-i\boldsymbol{kr}} w(\boldsymbol{r}), \tag{13}$$

and the *structure factor*

$$S(\boldsymbol{k}) = \frac{1}{N} \sum_{\boldsymbol{R}} e^{-i\boldsymbol{kR}} \tag{14}$$

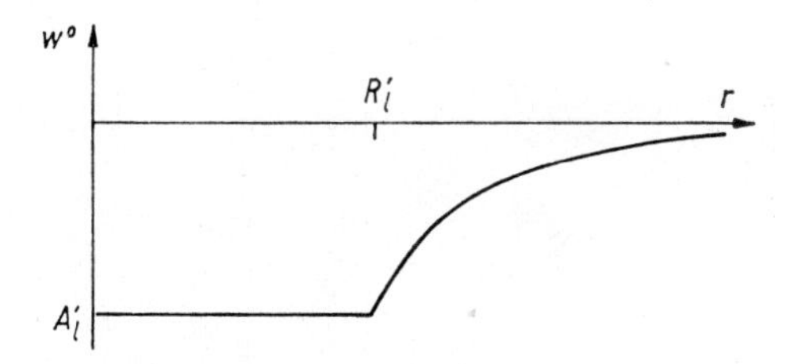

Fig. 2.2-3. Optimized pseudopotential after Shaw for a single ion

characterizing the arrangement of the N atoms in the crystal. The screened model potential $w(\boldsymbol{k})$ results from the potential (10) of the "bare" ion $w^0(\boldsymbol{k})$ and the *dielectric function* $\varepsilon^{(e)}(\boldsymbol{k})$ from (5.2-22) describing the screening by the gas of nearly free electrons,

$$w(\boldsymbol{k}) = w^0(\boldsymbol{k})/\varepsilon^{(e)}(\boldsymbol{k}). \tag{15}$$

In (12) and (15) the complete separation of properties of free ions (w^0), their arrangement (S), and the screening in the metal ($\varepsilon^{(e)}$) is remarkable.

The smoother the model potential $W(\boldsymbol{r})$ behaves, the smaller the Fourier transform $W(\boldsymbol{k})$ for large k (i.e. for small wave length). This fact is used in the construction of the "optimized pseudopotential" according to Shaw (68). In this procedure, the potential step at R_l' is avoided by choosing R_l' in (10) according to $A_l'(\varepsilon) = 1/R_l'$. Corrections of (15) as a consequence of the non-

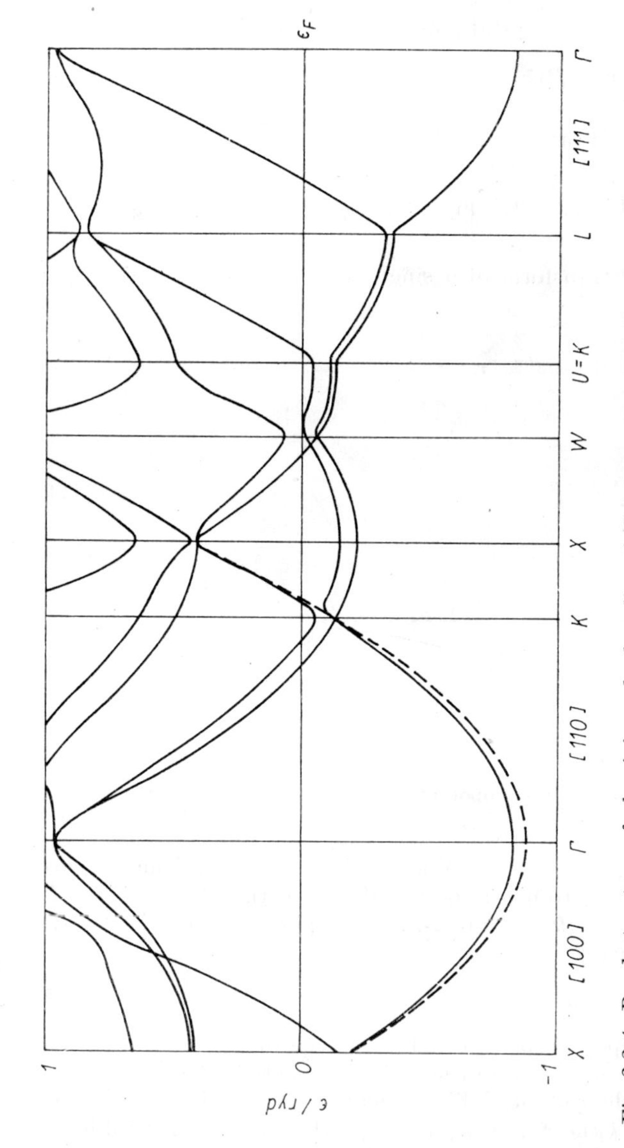

Fig. 2.2-4. Band structure of aluminium; broken line: behaviour of free electrons ($\varepsilon_F = 0.868$ *ryd*); lattice constant $a = 7.60a_0$, self-consistent LCAO calculation with the Hedin-Lundqvist exchange-correlation potential (Eschrig 82 a)

locality of the potential are of only minor influence on band structure calculations and, therefore, are only mentioned here (this is not so for the calculation of derived quantities, such as phonon spectra).

The calculated band structure of aluminium is shown in fig. 2.2-4. As in fig. 1.2-7, the eigenvalues are drawn along straight lines between points of the Brillouin zone in fig. 1.2-5 with high symmetry. The broken line gives the energy of free electrons. At the point Γ in the middle of the Brillouin zone, we observe the parabolae for nearly free electrons (along lines $\overline{\Gamma X}$ and $\overline{\Gamma K}$). The line $\overline{XW}$ at the surface of the Brillouin zone is formed for free electrons, also from a piece of a parabola. For electrons in a constant potential, these pieces are degenerated according to (1.2-22). The lattice potential breaks this degeneration. According to the considerations in [1.2] the band splitting amounts $2W_G$. From the width of the splitting on the line $\overline{XW}$, it may be seen that the Fourier component W_G of the potential is small compared with the unperturbed energy $\varepsilon = \hbar^2k^2/2m$. This consideration proves the validity of a perturbation treatment. Near the Γ point, the parabola-like behaviour is perturbed by a second-order correction. Only in the immediate vicinity of the zone boundary is the perturbation of first order.

The method of model or pseudopotentials illustrated in this chapter for band structure calculations represents an extremely efficient tool for the quantitative calculation of many physical properties, because it enables (i) treatment of the influence of the crystal potential via a perturbation theory of second order starting from free electrons, and (ii) many-body effects to be taken into account, which are important for evaluation of properties such as the total energy of the system via the theory of the interacting electron gas (by means of the dielectric function in (15), see section [5.2]). In this manner quantitative results on stable lattice structure and elastic moduli [5.3.1], spectrum of lattice vibrations [5.3.2], and transport properties such as electrical resistivity, may be obtained.

2.2.4 Norm-Conserving Pseudopotentials

Recent progress in the use of pseudopotentials rests on the concept of norm-conservation, which is described briefly in the following, along the lines of Appendix A.2 of Hafner (87).

Returning once more to [2.2.2 and 2.2.3] and summarizing these sections, the idea of pseudopotentials or valence pseudo-orbitals (inside the core region $r < R_c$) can be expressed by the following conditions:

(1) true orbitals and pseudo-orbitals have the same eigenvalues;
(2) the radial part of the pseudo-orbital is nodeless and identical to the true valence orbital outside the core region ($r > R_c$);
(3) the first (and second) derivative of the pseudo-orbital is matched to the exact values at R_c.

These conditions lead to non-norm-conserving pseudopotentials. Haman et al. (79) introduced additionally the norm-conservation condition

(4) the pseudocharge contained in a sphere of radius R_c is identical to the real charge in that sphere.

Applying this condition, norm-conserving pseudopotentials result. They are thought to optimize the transferability of the pseudopotential, because the scattering properties of the pseudopotential and of the full potential have the same energy variation to first order when transferred to another configuration, e.g. from the free atom to an ion in the metal.

There are several possibilities for constructing norm-conserving pseudopotentials. Whereas Hamann et al. (7) start with atomic potentials, Kerker (80) starts with the true valence orbital. In contrast to the Phillips-Kleinman type pseudopotentials (being strongly repulsive or oscillatory in the core region), the norm-conserving potentials are smooth. This improves their usefulness in band structure calculations, which use plane waves as the basis set. Ihm et al. (79) and Nielsen and Martin (85) applied this type of pseudopotential to calculate total energies and stress (elastic properties of Si, Ge, GaAs), respectively. Although norm-conserving pseudopotentials have attractive features, their use in perturbation calculations is not free of problems.

2.3 Multiple Scattering Picture

2.3.1 Muffin-Tin Potential

The concrete form of the band structure is determined by the behaviour of the potential. In metals with nearly free electrons, the potential can be described in very good approximation by the muffin-tin model (fig. 2.3-1). In the vicinity of the ions the potential for the conduction electrons is determined by the charge of the nuclei and the strongly localized core electrons. Therefore, the potential is considered to be spherically symmetric inside the muffin-tin (MT) spheres marked in fig. 2.3-1. In this region, it varies very strongly and is determined by deep boxes as for free ions. In the intermediate region between the MT spheres (MT floor), the potential is assumed to be constant. This is a good approximation for metals because of the dense packing of atoms and the high symmetry around them in contrast with the open structures of semiconductors. Thus, the approximation takes into account both the behaviour of the potential in the immediate vicinity of the ions and the existence of nearly free conduction electrons in metals.

The MT approximation enables a fast band structure calculation also in those cases where the potential contrary to [2.2] cannot be considered to be small. This situation arises, for example, in transition metals. For a recent review on multiple scattering theory and its applications see Weinberger (90).

2.3.2 Scattering Representation

In the interstitial region of the MT potential, the motion of an electron can be described by plane waves, scattered permanently on the spherical potential boxes. Such a multiple scattering process is sketched in fig. 2.3-1. The summation of all possible scattering events is possible making use of the translational symmetry of the arrangement of the single potential boxes. It yields the band structure $\varepsilon(\boldsymbol{k})$.

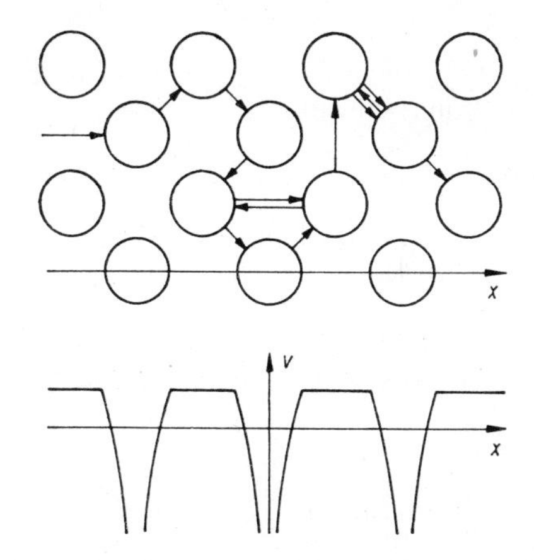

Fig. 2.3-1. MT potential. Inside the spheres around the ionic sites the potential is spherical, outside it is constant. The arrows denote the path of an electron in a multiple-scattering process. The lower part shows the potential along the line x

The scattering of plane waves on a spherically symmetric potential is a problem well known from quantum mechanics. It requires the solution of the radial Schroedinger equation

$$\left(-\frac{1}{r}\frac{\mathrm{d}^2}{\mathrm{d}r^2}r - \varkappa^2 + v_{\mathrm{eff}}(r)\right) R_l(r, \varkappa) = 0 \tag{1}$$

with

$$\varepsilon = \frac{\hbar^2}{2m}\varkappa^2; \quad v(r) = \frac{\hbar^2}{2m}v'(r), \quad v_{\mathrm{eff}}(r) = v'(r) + \frac{l(l+1)}{r^2} \tag{2}$$

where $l(l+1)/r^2$ represents the potential of the centrifugal force in a state with angular momentum l. The solutions $R_l(r, \varkappa)$ near $r = 0$ with asymptotic behaviour

$$R_l(r, \varkappa) \sim r^l \quad \text{for} \quad r \to 0 \tag{3}$$

are of special interest. For a vanishing potential $v' = 0$, the two fundamental solutions of (1) are given by the spherical Bessel and Neumann functions j_l and n_l, respectively, with asymptotic behaviour:

	$r \to 0$	$r \to \infty$	
$\mathrm{j}_l(\varkappa r) \sim$	r^l	$\dfrac{1}{\varkappa r} \sin(\varkappa r - \pi l/2)$	(4)
$\mathrm{n}_l(\varkappa r) \sim$	$-r^{-l-1}$	$-\dfrac{1}{\varkappa r} \cos(\varkappa r - \pi l)/2)$	

Outside R_{MT}, the radial solution R_l may be superposed by j_l and n_l,

$$R_l(r, \varkappa) = \mathrm{j}_l(\varkappa r) \cos \eta_l(\varkappa) - \mathrm{n}_l(\varkappa r) \sin \eta_l(\varkappa) . \tag{5}$$

Any information on the wave function, and along with it the scattering properties of the potential, is contained in the energy dependent scattering phase shifts $\eta_l(\varkappa)$ for angular momentum l.

The meaning of $\eta_l(\varkappa)$ as phase shifts of the unperturbed waves $\mathrm{j}_l(\varkappa r)$ becomes obvious in the following form of the asymptotic representation

$$R_l(r, \varkappa) \sim \frac{1}{\varkappa r} \sin(\varkappa r + \eta_l - \pi l/2) \quad \text{for} \quad r \to \infty . \tag{6}$$

The situation is presented in fig. 2.3-2 once more in an obvious way. The radial wave functions for $l = 0$ are shown both in the potential $v(r)$ and in the con-

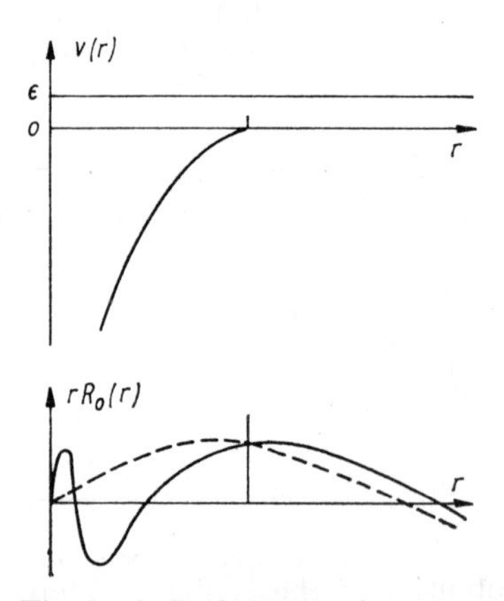

Fig. 2.3-2. Radial wave functions $rR_0(r)$ at energy ε in the potential well $v(r)$ with two nodes within the well (solid line) and in a constant potential $v = 0$ (broken line). Outside the well ($r > R$) they differ only by a phase shift η_0

stant potential. Inside the MT sphere, they are very different, but outside they show apart from the phase shift η_0 the same oscillating behaviour.[1]) Since the MT spheres do not overlap, for their interaction only these phase shifts are relevant. Therefore, in this picture the atoms may be treated as black boxes, characterized by their scattering phase shifts $\eta_l(\varkappa)$ as functions of energy. The phase shifts are calculated by the numerical integration of the ordinary differential equation (1) from $r = 0$ to R_{MT}, and the function is fitted to the representation (5). Since the arbitrary normalization is given by (5), the fit to the logarithmic derivative

$$L_l(\varkappa) \equiv \frac{R_l'(R_{\text{MT}}, \varkappa)}{R_l(R_{\text{MT}}, \varkappa)} = \varkappa \frac{j_l'(\varkappa R_{\text{MT}}) \cos \eta_l(\varkappa) - n_l'(\varkappa R_{\text{MT}}) \sin \eta_l(\varkappa)}{j_l(\varkappa R_{\text{MT}}) \cos \eta_l(\varkappa) - n_l(\varkappa R_{\text{MT}}) \sin \eta_l(\varkappa)} \tag{7}$$

is sufficient. The primes on j and n denote derivatives with respect to the arguments of the functions.

2.3.3 Behaviour of Potential and Phase Shifts in Metals

For simple metals in the interesting energy region of the conduction bands, the phase shifts have values near multiples of π. Since a change in phase shift by π implies only a non-observable shift of the outer part of the wave function by exactly one node, it means that the electrons pass the scattering centre nearly unperturbed, even though the real potential is by no means weak! The

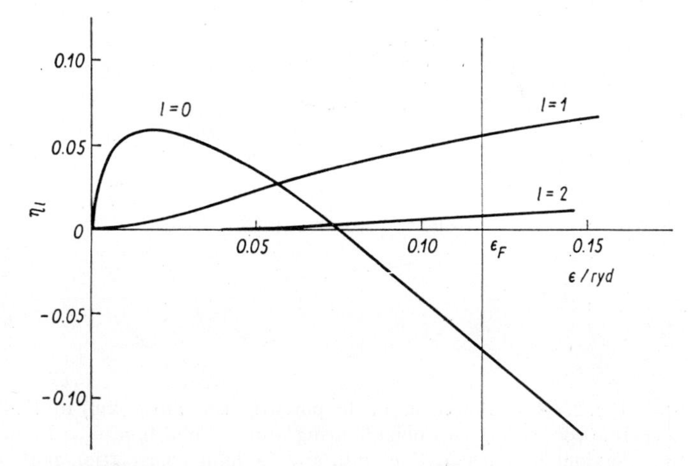

Fig. 2.3-3. Calculated scattering phase shifts η_l for sodium (after Paasch and Woittennek 73) as a function of energy for the angular momenta $l = 0, 1, 2$.

[1]) For $l = 0$, j_0 and n_0 have the simple form $j_0 = (\varkappa r)^{-1} \sin(\varkappa r)$ and $n_0 = -(\varkappa r)^{-1} \cos(\varkappa r)$, respectively

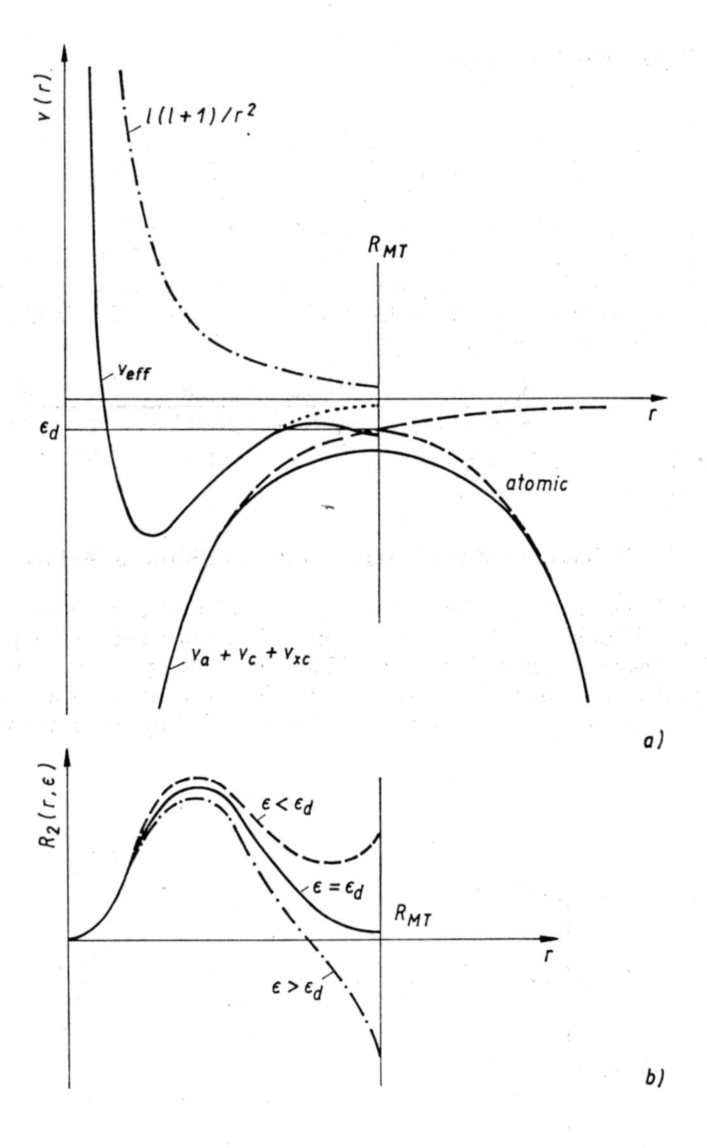

Fig. 2.3-4. a) Behaviour of the potential (in units $\hbar^2/2m$) for d-states on the line connecting two neighbouring atoms. Solid line: v_{eff}; broken line: behaviour of nuclear, Coulomb, and exchange-correlation potential $v' = v_a + v_c + v_{xc}$ for isolated atoms; lower solid line: superposition of v' for two neighbouring atoms. v_{eff} denotes the sum of v' and of the centrifugal potential $l(l+1)/r^2$ (dash-dotted); dotted: v_{eff} for isolated atoms.
b) Behaviour of the radial wave function $R_2(r, \varepsilon)$ of a d-state in the potential a) at the resonance energy ε_d and wave functions for a somewhat larger (dash-dotted line) and smaller (dashed line) energy, respectively

reduction in phase shift by a term $n\pi$ to a smaller quantity η' just means removing the nodes of the true wave function near the ionic cores. This is a somewhat different representation of the smallness of the pseudo or model potential treated in the last chapter. As an example for simple metals in fig. 2.3-3 the calculated scattering phase shifts for Na are given. The values at the Fermi energy are especially interesting because they determine the shape of the Fermi surface and the transport properties. The smallness of the reduced phase shifts η' explains the fact that in simple metals the valence electrons behave like nearly free electrons.

In transition and noble metals, the d-electrons are essential. The d-levels in the isolated atoms are situated in the energy region of the s- and p-levels of the following shell. However, the wave functions of the d-electrons are much more localized than those of the s- and p-electrons. Passing to a crystal, this is the reason for the smaller overlap of wave functions of neighbouring atoms and, therefore, for narrower bands of the d-electrons situated within the band of the nearly free electrons resulting from atomic s- and p-levels. The reason

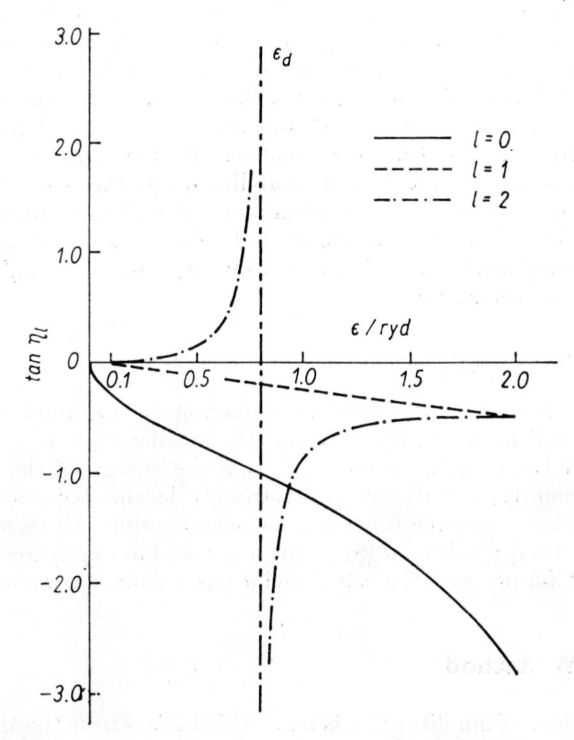

Fig. 2.3-5. Tangent of scattering phase shifts $\tan \eta_l$ for $l = 0, 1, 2$ for vanadium as a function of energy (Urwank 74)

of the quantitative difference is given by the increasing influence of the potential of the centrifugal force $\sim l(l+1)/r^2$.

To explain this fact, in fig. 2.3-4, the behaviour of the potential on the line connecting the centres of two neighbouring atoms is shown. The MT potential results from superposition of the atomic potentials showing a broad maximum between the atoms. In classical mechanics it marks the boundary between free and bound states and, therefore, it is a measure for the bottom of the conduction band. For d-electrons, we have to add in each MT sphere the potential of the centrifugal force. In consequence, we obtain a potential well capturing the electron. Because of the potential threshold produced by the centrifugal potential, a d-electron with the marked energy ε_d can tunnel to the neighbouring well with a small probability only. Therefore, the wave function decreases exponentially in the region of the potential barrier as is shown in fig. 2.3-4b). For energies deviating only a little from ε_d the exponentially increasing part of the fundamental solution becomes important. At the beginning of the barrier, the differences between the wave function are still small, but increase quickly as shown in fig. 2.3-4b), inducing a strong qualitative change in attached phase shift in this energy region. Near ε_d, the tangent of the wave function nearly vanishes, but for lower (higher) energies it shows considerable positive (negative) values. In this case, the plane wave in the constant potential region outside the MT spheres has nodes near the MT radius. In the language of scattering theory, the strongly energy dependent behaviour described is denoted as a resonance. At the resonance energy ε_d, the particle may be found with high probability inside the potential well, as may be seen from the ratio of the internal and external wave functions. The scattering properties — in the case considered, the scattering phase shifts — change very quickly within the resonance width Γ_d from values η' near zero over $\pi/2$ at ε_d to values near π:

$$\tan \eta_2 = \frac{\Gamma_d}{\varepsilon_d - \varepsilon} + \tan \eta_2'; \quad \tan \eta_2' \ll 1. \tag{8}$$

The larger the life time τ of the quasi-stationary state at ε_d and the potential barrier determining its decay, respectively, the smaller $\Gamma_d = \hbar/\tau$. For an infinitely extended barrier ($R_{\mathrm{MT}} \to \infty$), ε_d equals the energy of the atomic d-electron. Consequently, the existence of strongly localized d-electrons with only weakly overlapping wave functions at neighbouring lattice sites corresponds to pronounced resonances (8) of the d-phase shifts η_2 in the scattering picture. Figure 2.3-5 presents the calculated d-phase shifts for vanadium.

2.4 APW Method

In the APW method, a simplification is used which arises from the MT approximation by describing the wave function in the interstitial region of the potential as a linear combination of plane waves augmented at the surface of the MT spheres continuously to the spherically symmetric solutions inside the MT spheres (APW: Augmented Plane Wave):

$$\Phi_{\boldsymbol{k}}(\boldsymbol{r}) = \begin{cases} \mathrm{e}^{i\boldsymbol{k}\boldsymbol{r}}, \\ \sum_L i^l a_L R_l(r, \varkappa)\, \mathcal{Y}_L(\boldsymbol{r}) \end{cases} \quad (r \gtrless R_{\mathrm{MT}}). \tag{1}$$

L is short hand for the quantum numbers of angular momentum l and its z-component m. $\mathcal{Y}_L(\boldsymbol{r})$ denotes the complex spherical harmonics depending on the variables ϑ and φ (r, ϑ, φ are spherical coordinates). The coefficients a_L may be obtained using the auxiliary formula

$$\mathrm{e}^{i\boldsymbol{k}\boldsymbol{r}} = 4\pi \sum_L i^l \mathrm{j}_l(kr)\, \mathcal{Y}_L(\boldsymbol{r})\, \mathcal{Y}_L^*(\boldsymbol{k}) \tag{2}$$

from the continuity requirement of (1) at $r = R_{\mathrm{MT}}$:

$$a_L = 4\pi \frac{\mathrm{j}_l(kR_{\mathrm{MT}})}{R_l(R_{\mathrm{MT}}, \varkappa)} \mathcal{Y}_L^*(\boldsymbol{k}). \tag{3}$$

In analogy to the OPW's in (2.2-4), the solutions $\psi_{\boldsymbol{k}}(\boldsymbol{r})$ of the Schroedinger equation (1.2-1) is described by the linear combination of the APW's (1) with correct translational symmetry:

$$\psi_{\boldsymbol{k}}(\boldsymbol{r}) = \sum_{\boldsymbol{G}} \alpha_{\boldsymbol{k}+\boldsymbol{G}} \Phi_{\boldsymbol{k}+\boldsymbol{G}}(\boldsymbol{r}). \tag{4}$$

The coefficients $\alpha_{\boldsymbol{k}+\boldsymbol{G}}$ are determined again via the extremal property of the total energy by variation.

Because of the discontinuity of the derivative of $\Phi_{\boldsymbol{k}}$ at R_{MT}, the expectation value of the kinetic energy (in units $\hbar^2/2m$) must be used in the form $\int \mathrm{d}^3r \left(\frac{\partial\psi}{\partial\boldsymbol{r}}\right)^* \frac{\partial\psi}{\partial\boldsymbol{r}}$ instead of the usual expression $\int \mathrm{d}^3r\psi^* \left(-\frac{\partial^2}{\partial\boldsymbol{r}^2}\psi\right)$. With it, the quantity $(\overline{H} - \varkappa^2)$ is expressed for one Wigner-Seitz cell (all the cells yield the same result because of the translational symmetry) as

$$\begin{aligned} &\int_{\mathrm{WS}} \mathrm{d}^3r \left\{\frac{\partial\psi^*}{\partial\boldsymbol{r}} \frac{\partial\psi}{\partial\boldsymbol{r}} + \left(v'(\boldsymbol{r}) - \varkappa^2\right) \psi^*\psi\right\} \\ &= \left(\int_{\mathrm{MT}} \mathrm{d}^3r + \int_{\mathrm{res}} \mathrm{d}^3r\right) \left\{\psi^* \left(-\frac{\partial^2}{\partial\boldsymbol{r}^2} + v(\boldsymbol{r}) - \varkappa^2\right) \psi\right\} \\ &\quad + \oiint_{(\mathrm{MT})} \mathrm{d}\boldsymbol{S}\psi^* \frac{\partial}{\partial\boldsymbol{r}} (\psi_{\mathrm{in}} - \psi_{\mathrm{out}}). \end{aligned} \tag{5}$$

Above all, the volume integral on the elementary cell has been divided into integrals on the MT sphere and the residual (interstitial) region (compare fig. 2.4-1) and after that the integral

$$\int_{\mathrm{WS}} \mathrm{d}^3r \, \frac{\partial \psi^*}{\partial \boldsymbol{r}} \frac{\partial \psi}{\partial \boldsymbol{r}} = -\int_{\mathrm{WS}} \mathrm{d}^3r \psi^* \frac{\partial^2}{\partial \boldsymbol{r}^2} \psi + \oiint_{(\mathrm{WS})} \mathrm{d}\boldsymbol{S} \psi^* \frac{\partial \psi}{\partial \boldsymbol{r}} \tag{6}$$

was transformed by means of the Gaussian theorem into the usual form and a surface integral on the considered Wigner-Seitz cell (WS). ψ_{in} and ψ_{out} denote the solutions inside and outside the MT sphere, respectively. The integral on the surface of the elementary cell vanishes because of the translational symmetry of the wave function $\psi_{\boldsymbol{k}}$. The first integral in (5) on MT also vanishes, since in this region the ansatz (1) fulfils the Schroedinger equation. The integral on the residual volume with a constant potential may easily be evaluated because of the plane waves in that region.[1])

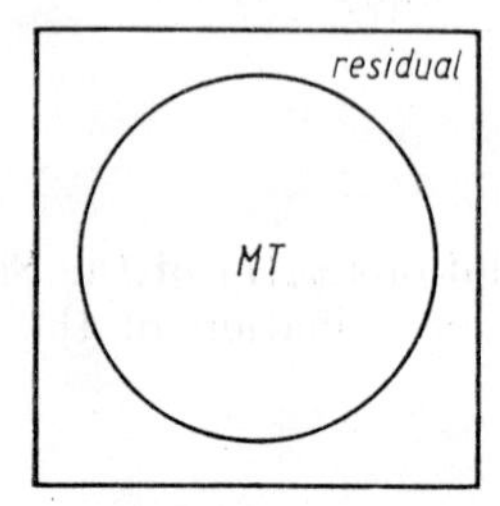

Fig. 2.4-1. Region of integration in (2.4-5). The elementary cell (schematically represented as a square) is divided into the MT sphere and the remaining region

The surface integral in (5) vanishes if the $\alpha_{\boldsymbol{k}+\boldsymbol{G}}$ in (4) are chosen so that $\partial\psi/\partial\boldsymbol{r}$ becomes continuous. This is the case for solution (7) of the problem. Because of the spherical symmetry, the surface integral in (5) may be evaluated easily. Using (1), it yields

$$\sim \mathrm{j}_l(|\boldsymbol{k} + \boldsymbol{G}| \, R_{\mathrm{MT}}) \, \mathrm{j}_l(|\boldsymbol{k} + \boldsymbol{G}'| \, R_{\mathrm{MT}}) \left\{ \frac{R_l'(R_{\mathrm{MT}}, \varkappa)}{R_l(R_{\mathrm{MT}}, \varkappa)} - |\boldsymbol{k} + \boldsymbol{G}'| \frac{\mathrm{j}_l'(|\boldsymbol{k} + \boldsymbol{G}'| \, R_{\mathrm{MT}})}{\mathrm{j}_l(|\boldsymbol{k} + \boldsymbol{G}'| \, R_{\mathrm{MT}})} \right\}.$$

Finally, the variation (5) with respect to the coefficients yields the *APW equations* (Slater 37, Loucks 67)

$$\sum_{\boldsymbol{G}'} \{[(\boldsymbol{k} + \boldsymbol{G})^2 - \varkappa^2] \, \delta_{\boldsymbol{G}\boldsymbol{G}'} + \Gamma_{\boldsymbol{G}\boldsymbol{G}'}(\varkappa)\} \, \alpha_{\boldsymbol{k}+\boldsymbol{G}'} = 0 \tag{7}$$

[1]) Note that ψ in (4) does not represent a solution of the Schroedinger equation in that region, but is superposed by plane waves belonging to different energies $(\boldsymbol{k} + \boldsymbol{G})^2$.

with

$$\Gamma_{GG'}(\varkappa) = \frac{4\pi R_{\mathrm{MT}}^2}{\Omega_0} \left\{ -[(\boldsymbol{k}+\boldsymbol{G})^2 - \varkappa^2] \frac{\mathrm{j}_l(|\boldsymbol{G}-\boldsymbol{G}'|\, R_{\mathrm{MT}})}{|\boldsymbol{G}-\boldsymbol{G}'|} \right.$$

$$+ \sum_{l=0}^{\infty} (2l+1)\, P_l(\zeta)\, \mathrm{j}_l(|\boldsymbol{k}+\boldsymbol{G}|\, R_{\mathrm{MT}})\, \mathrm{j}_l(|\boldsymbol{k}+\boldsymbol{G}'|\, R_{\mathrm{MT}})$$

$$\left. \times \left[\frac{R_l'(R_{\mathrm{MT}}, \varkappa)}{R_l(R_{\mathrm{MT}}, \varkappa)} - |\boldsymbol{k}+\boldsymbol{G}| \frac{\mathrm{j}_l'(|\boldsymbol{k}+\boldsymbol{G}|\, R_{\mathrm{MT}})}{\mathrm{j}_l(|\boldsymbol{k}+\boldsymbol{G}|\, R_{\mathrm{MT}})} \right] \right\}. \tag{8}$$

$P_l(\zeta)$ denotes the Legendre polynomials with $\zeta = \cos \sphericalangle (\boldsymbol{k}+\boldsymbol{G}, \boldsymbol{k}+\boldsymbol{G}')$. The expression $\Gamma_{GG'}(\varkappa)$ in (7) is reminiscent of the matrix elements of a pseudopotential. However, in the case of transition metals it is not small. Moreover, it is very strongly energy-dependent, and even exhibits poles with respect to $\varkappa$ at zeros of the function $R_l(R_{\mathrm{MT}}, \varkappa)$.

Solving eqn. (7), one can fix the value of $\boldsymbol{k}$ and seek the zeros of the determinant in parentheses with respect to $\varkappa$, or one can fix $\varkappa$ and the direction of $\boldsymbol{k}$ and seek k solving (7). In both cases, the numerical expense is considerable firstly because of the large number of APW's to be considered in (7) (approximately 50 per atom in the elementary cell); secondly because of the sums over l in (8) with about 12 terms; and finally because of the strong energy-dependency of the matrix elements. During each step of an iterative solution, every matrix element has to be calculated anew.

2.5 KKR Method

2.5.1 Scattering Representation and Green Function

The KKR method (named after Korringa 47, Kohn and Rostoker 54) employs consequently the multiple scattering representation in the MT potential. To this aim, a representation of the problem equivalent to the Schroedinger equation (1.2-1) using the homogeneous integral equation

$$\psi(\boldsymbol{r}) = \int \mathrm{d}^3r'\, G(\boldsymbol{r}-\boldsymbol{r}')\, v'(\boldsymbol{r}')\, \psi(\boldsymbol{r}') \tag{1}$$

with the Green function $G(\boldsymbol{r}-\boldsymbol{r}')$ of free particles

$$G(\boldsymbol{r}-\boldsymbol{r}') = -\frac{\mathrm{e}^{i\varkappa|\boldsymbol{r}-\boldsymbol{r}'|}}{4\pi\, |\boldsymbol{r}-\boldsymbol{r}'|} \tag{2}$$

with retarded waves or with standing waves

$$G^{\mathrm{S}}(\boldsymbol{r}-\boldsymbol{r}') = -\frac{\cos \varkappa\, |\boldsymbol{r}-\boldsymbol{r}'|}{4\pi\, |\boldsymbol{r}-\boldsymbol{r}'|} \tag{2'}$$

is advantageous. Both (2) and (2′) are solutions of the differential equation

$$\left(-\frac{\partial^2}{\partial \boldsymbol{r}^2} - \varkappa^2\right) G(\boldsymbol{r} - \boldsymbol{r}') = -\delta(\boldsymbol{r} - \boldsymbol{r}'). \tag{3}$$

Since $v'(\boldsymbol{r})$ in (1) represents a sum on isolated MT potentials on the right hand side of (1) with (2), retarded waves from each potential well are contained. Using the translational symmetry of the wave function via the Bloch theorem (1.2-5) for each MT sphere the same integral is obtained, but with different phase factors $\exp(i\boldsymbol{kR})$,

$$\psi_{\boldsymbol{k}}(\boldsymbol{r}) = \int_{\mathrm{WS}} \mathrm{d}^3r' G_{\boldsymbol{k}}(\boldsymbol{r} - \boldsymbol{r}')\, v'(\boldsymbol{r}')\, \psi_{\boldsymbol{k}}(\boldsymbol{r}') \tag{4}$$

with

$$G_{\boldsymbol{k}}(\boldsymbol{r}) = \sum_{\boldsymbol{R}} G(\boldsymbol{r} - \boldsymbol{R})\, \mathrm{e}^{i\boldsymbol{kR}}. \tag{5}$$

$G_{\boldsymbol{k}}$ contains the Bloch boundary condition and, therefore, information on the structure. The term for $\boldsymbol{R} = 0$ in (5) exhibits a singularity at $\boldsymbol{r} = 0$ as may be seen from (2) or (2′). According to (3), $G(\boldsymbol{r} - \boldsymbol{r}')$ may be expanded in linear combinations of solutions of the homogeneous Schroedinger equation in a constant potential (except the case $\boldsymbol{r} = \boldsymbol{r}'$). In spherical coordinates, these fundamental solutions are given by the spherical Bessel and Neumann functions[1]):

$$G(\boldsymbol{r} - \boldsymbol{r}') = \sum_L Y_L(\boldsymbol{r})\, Y_L(\boldsymbol{r}')\, \mathrm{j}_l(\varkappa r)\, \big(\mathrm{n}_l(\varkappa r') - i\mathrm{j}_l(\varkappa r')\big) \quad \text{for} \quad r \leqq r' \tag{6}$$

and

$$G^{\mathrm{S}}(\boldsymbol{r} - \boldsymbol{r}') = \sum_L Y_L(\boldsymbol{r})\, Y_L(\boldsymbol{r}')\, \mathrm{j}_l(\varkappa r)\, \mathrm{n}_l(\varkappa r') \quad \text{for} \quad r \leqq r', \tag{6'}$$

respectively.

The discontinuity of the derivatives of G and G^{S} at $r = r'$ yields exactly the singularity in (3).

The terms in (5) for $\boldsymbol{R} \neq 0$ may be represented by the regular fundamental solutions. Representing (5) in spherical coordinates inside the Wigner-Seitz cell considered, the form

$$G_{\boldsymbol{k}}(\boldsymbol{r} - \boldsymbol{r}') = \sum_{LL'} \mathrm{j}_l(\varkappa r)\, \{i^{l-l'} B_{LL'} \mathrm{j}_{l'}(\varkappa r') + \delta_{LL'}[\mathrm{n}_l(\varkappa r') - i\mathrm{j}_l(\varkappa r')]\}\, Y_L(\boldsymbol{r})\, Y_{L'}(\boldsymbol{r}') \quad \text{for} \quad r \leqq r' \tag{7}$$

[1]) As opposed to (2.4-2), here the use of real linear combinations Y_L of the complex conjugated spherical harmonics $\mathcal{Y}_{lm}$ and $\mathcal{Y}_{l,-m}$, respectively, is of advantage.

is obtained, which is well suited for forthcoming calculations. Likewise, it is possible to represent the Green function by plane waves with a suitable translational symmetry:

$$G_{\boldsymbol{k}}(\boldsymbol{r} - \boldsymbol{r}') = -\sum_{\boldsymbol{G}} \frac{e^{i(\boldsymbol{k}+\boldsymbol{G})(\boldsymbol{r}-\boldsymbol{r}')}}{(\boldsymbol{k} + \boldsymbol{G})^2 - \varkappa^2 - i\delta}, \quad \delta \gtrsim 0. \tag{8}$$

It is easily proved, by putting (8) into (3), that $G_{\boldsymbol{k}}$ is a solution of (3) also obeying the Bloch condition well. The positive infinitesimal quantity δ in the denominator of (8) guarantees that (8) fulfils as does (2) the condition for retarded waves. Comparing (8) with (7) and using the expansion theorem (2.4-2) for plane waves at R_{MT}, the following representation for the energy dependent matrix B is obtained:

$$\begin{aligned} B_{LL'} &= \sum_{\boldsymbol{G}} \frac{\Phi^{\boldsymbol{k}}_{L\boldsymbol{G}} \Phi^{\boldsymbol{k}*}_{L'\boldsymbol{G}}}{\varkappa^2 + i\delta - (\boldsymbol{k} + \boldsymbol{G})^2} - \delta_{LL'} \left(\frac{\mathrm{n}_l(\varkappa R_{\mathrm{MT}})}{\mathrm{j}_l(\varkappa R_{\mathrm{MT}})} - i \right) \\ &= B^S_{LL'} + i\delta_{LL'} \end{aligned} \tag{9}$$

with

$$\Phi^{\boldsymbol{k}}_{L\boldsymbol{G}} = -\frac{4\pi i^l \mathrm{j}_l(|\boldsymbol{k} + \boldsymbol{G}|\, R_{\mathrm{MT}})\, Y_L(\boldsymbol{k} + \boldsymbol{G})}{(\Omega_0 \varkappa)^{1/2}\, \mathrm{j}_l(\varkappa R_{\mathrm{MT}})}. \tag{10}$$

2.5.2 Wave Functions and KKR Equations

In analogy to (2.4-1), we represent the wave function inside the considered elementary cell by solutions of the radial Schroedinger equation:

$$\psi_{\boldsymbol{k}} = \sum_L i^l C_L(\boldsymbol{k})\, R_l(r, \varkappa)\, Y_L(\boldsymbol{r}). \tag{11}$$

Since the potential in (4) vanishes outside the MT sphere we need $\psi_{\boldsymbol{k}}$ only for $r \leq R_{\mathrm{MT}}$. However, the expression (11) is valid in the total elementary cell if the coefficients $C_L(\boldsymbol{k})$ fulfil the solubility condition (16).

Using (7) and (11), the evaluation of (4) is simple. The appearing volume integrals of type

$$J_{LL'} = \int \mathrm{d}^3 r \mathrm{j}_l(\varkappa r)\, Y^*_L(\boldsymbol{r})\, v'(r)\, R_{l'}(r, \varkappa)\, Y_{L'}(\boldsymbol{r}) \tag{12}$$

may be evaluated easily because of the orthogonality of the spherical harmonics

$$\int \mathrm{d}\vartheta \sin\vartheta\, \mathrm{d}\varphi\, Y^*_L(\boldsymbol{r})\, Y_{L'}(\boldsymbol{r}) = \delta_{LL'} \tag{13}$$

as is shown in the following steps:

$$J_{LL'} = \delta_{LL'} \int_0^{R_{\mathrm{MT}}} \mathrm{d}r r^2 \mathrm{j}_l(\varkappa r)\, v'(r)\, R_l(r, \varkappa)$$

$$= \delta_{LL'} \int_0^{R_{\mathrm{MT}}} \mathrm{d}r r^2 \mathrm{j}_l(\varkappa r) \left(\frac{1}{r}\frac{\mathrm{d}^2}{\mathrm{d}r^2} r + \varkappa^2 - \frac{l(l+1)}{r^2}\right) R_l(r, \varkappa)$$

$$= \delta_{LL'} \left[r \mathrm{j}_l(\varkappa r) \frac{\mathrm{d}}{\mathrm{d}r} \left(r R_l(r, \varkappa)\right) - r R_l(r, \varkappa) \frac{\mathrm{d}}{\mathrm{d}r} \left(r \mathrm{j}_l(\varkappa r)\right)\right]_0^{R_{\mathrm{MT}}}$$

$$= -\frac{1}{\varkappa} \sin \eta_l \delta_{LL'}. \tag{14}$$

Here R_l represents a solution of the radial Schroedinger equation (2.3-1 and 2) and j_l fulfils the same equation with $v' = 0$. The last line is obtained using (2.3-5) and the Wronski relation for j_l and n_l

$$\mathrm{j}_l(\varkappa r) \frac{\mathrm{d}}{\mathrm{d}r} \mathrm{n}_l(\varkappa r) - \mathrm{n}_l(\varkappa r) \frac{\mathrm{d}}{\mathrm{d}r} \mathrm{j}_l(\varkappa r) = \frac{1}{\varkappa r^2}. \tag{15}$$

The integrals $N_{LL'}$ following from (12) by replacing j_l by n_l yields analogously

$$N_{LL'} = -\frac{1}{\varkappa} \cos \eta_l \delta_{LL'}. \tag{14'}$$

With (14) and (14'), evaluation of (4) gives the homogeneous systems of linear equations for $C_L(\boldsymbol{k})$:

$$\sum_{L'} \left(B_{LL'}(\boldsymbol{k}, \varkappa) + \left(\cot \eta_l(\varkappa) - i\right) \delta_{LL'}\right) \sin \eta_l C_L(\boldsymbol{k}) = 0. \tag{16}$$

Together with expressions (9) and (10) for the structure constants, (16) represents the system of equations derived by KKR. Using standing waves (6') the imaginary unit i disappears because of compensation against the term i in (9). The solubility condition of (16) then reads as follows:

$$\det \|B_{LL'}^S(\boldsymbol{k}, \varkappa) + \delta_{LL'} \cot \eta_l(\varkappa)\| = 0. \tag{17}$$

For given $\boldsymbol{k}$, it is a non-linear equation for determination of the energy $\varkappa(\boldsymbol{k})$ as a function of $\boldsymbol{k}$. The separation of information on the arrangement of the atoms contained in the structure constants $B_{LL'}^S$ from information on the potential concentrated in the scattering phase shifts $\eta_l(\varkappa)$, is remarkable. The rank of (17) may be restricted to phase shifts η_l essentially non-vanishing. Consequently, for transition metals (17) proves to be a 9×9 matrix equation (for a given l the magnetic quantum number m may assume $(2l + 1)$ values; therefore, for the single bands we have the orientations of m given in parentheses: s (1), p (3), d (5)).

The main expense is connected with the calculation of the structure constants $B^S_{LL'}(\boldsymbol{k}, \varkappa)$. Both the lattice sum (9) in the reciprocal space and the lattice sum in real space derived directly from (5) converge badly, so that for their evaluation special techniques have to be applied. Moreover, because of (9), $B^S_{LL'}$ exhibits singularities just for the energies $\varkappa^2 = (\boldsymbol{k} + \boldsymbol{G})^2$ of free electrons. This demands an extremely fine-meshed net for the interpolation of determinant (17) as a function of energy.

As opposed to OPW and APW calculations, the KKR approach has the advantage that the dimension of the solution determinant is given by a two-parameter set (l, m), whereas in the OPW and APW methods, the matrix indices are given by the three-parameter set of vector $\boldsymbol{G}$. Information on structure and potential is separate. On the other hand, corrections on the MT potential (which are important in "open" structures like the diamond lattice) in the KKR approach cannot be taken into account as easily as in the APW method.

2.5.3. The General MT Potential

With the following idea it is tried to generalize the MT-concept: The true crystal potential $v(\boldsymbol{r})$ is considered as a sum of space filling cell potentials in such a way, that each cell potential is zero outside its WS-cell and agrees exactly with the crystal potential inside the cell. Thus the sum of all these cell potentials, which do not overlap or interpenetrate, reproduces completely the given crystal potential, simultaneously the MT-floor degenerates to infinitesimal thin skins along the surfaces of the WS-cells. This is the concept of general MT or space filling cell potentials. Each single cell can be described outside the bounding sphere which envelopes the WS cell by generalized (due to the non-spherically symmetric cell potential) phase shifts. In the crystal most of these spheres do not overlap (because the corresponding cells are far enough from each other), thus separating scattering properties and geometry for these cases. But the spheres of neighboured cells of course do overlap. This circumstance prevents a direct generalization of the usual *KKR* method with its typical separation between potential and structure to the case of space filling cell potentials (Ziesche 74). The question arises if it is nevertheless possible to overcome this problem and to develop a sophisticated calculational band structure method for that general case based on the concept of multiple-scattering theory.

Meanwhile a series of papers dealt with this question, the last ones are Badralexe and Freeman (88), Brown (88), Faulkner (88), Gonis (88), Molenaar (88, 89), Zeller (88). Besides this an attempt is made to use this concept especially for surfaces and interfaces (Gonis 86).

2.6 H-NFE-TB Method

2.6.1 Ziman Transformation

The KKR equations (2.5-17) are not very well suited to effective solution of the eigenvalue problem. This concerns not only the form of the structure constants $B^S_{LL'}$ given by (2.5-9) where poles at the energies of the free electrons

$\varkappa^2 = (\boldsymbol{k} + \boldsymbol{G})^2$ and the badly convergent sums on the lattice vectors are troublesome, but also the different behaviour according to (2.3-8) of those parts containing the scattering phase shifts via cot η_l. The latter are big for s- and p-contributions because of the smallness of the phase shifts η_0 and η_1, but on the other hand for $l = 2$ the cotangent just becomes zero in the interesting region near the resonance energy. The mentioned disadvantages may be avoided by a simple matrix manipulation based on the so-called Ziman transformation. Let us use the fact that (2.5-17) with (2.5-9) may be represented as

$$\det \|\sigma + A - \Phi(D - \varepsilon)^{-1} \Phi^+\| = 0. \tag{1}$$

σ and A are matrices with indices L and L' as in (2.5-17). D represents a diagonal matrix with elements $(\boldsymbol{k} + \boldsymbol{G})^2 \delta_{\boldsymbol{G}\boldsymbol{G}'}$. Corresponding to (2.5-10), Φ are rectangular matrices with indices L and $\boldsymbol{G}$.

The determinant (1) is equivalent to the extended determinant

$$\det \begin{Vmatrix} \sigma + A & \Phi \\ \Phi^+ & (D - \varepsilon) \end{Vmatrix} = 0 \tag{2}$$

(here the indices L and $\boldsymbol{G}$ have the same significance), since multiplication of the lower block of (2) with the matrix $\Phi(D - \varepsilon)^{-1}$ (from left) and the following subtraction from the upper block gives the determinant

$$\det \begin{Vmatrix} \sigma + A - \Phi(D - \varepsilon)^{-1} \Phi^+ & 0 \\ \Phi^+ & (D - \varepsilon) \end{Vmatrix} = 0. \tag{3}$$

Because of the zero-matrix in the upper right field, it decouples into the determinant (1) and in $\det \|D - \varepsilon\| = \prod_{\boldsymbol{G}} \left((\boldsymbol{k} + \boldsymbol{G})^2 - \varepsilon\right)$. With that, the inconvenient strong energy dependence of the structure constants $B^S_{LL'}$ via the poles in (1), can be avoided. The diagonal matrix $(D - \varepsilon)$ has exactly the form for free electrons. From the physical point of view, the extended matrix in (2) exhibits the desired expansion of the basic functions by the set of plane waves with correct translational symmetry. For practical calculations, it should be limited. However, this is not the case because of the strong coupling of the plane waves especially to the d-states (with $l = 2$) contained in $\Phi_{L\boldsymbol{G}}$ from (2.5-10). A limitation of the coupling is given for the contribution

$$\tilde{\Phi}_{L\boldsymbol{G}} = \Phi_{L\boldsymbol{G}} \exp\left[\left(-(\boldsymbol{k} + \boldsymbol{G})^2 + \varepsilon\right)/2\beta\right] \quad \text{für} \quad l = 2 \tag{4}$$

of $\Phi_{L\boldsymbol{G}}$, quickly decreasing with the convergency factor $\exp\left[\left(\varepsilon - (\boldsymbol{k} + \boldsymbol{G})^2\right)/2\beta\right]$.

This requires in (1) the separation of $\Phi(D - \varepsilon)^{-1} \Phi^+$ into a contribution $\tilde{\Phi}(D - \varepsilon)^{-1} \tilde{\Phi}^+$ which is used in the transformation (1) $\rightarrow$ (2) and into a residual part

$$A^{\boldsymbol{k}}_{LL'} = \sum_{\boldsymbol{G}} \frac{\Phi^{\boldsymbol{k}}_{L\boldsymbol{G}} \Phi^{\boldsymbol{k}*}_{L'\boldsymbol{G}}}{\varepsilon - (\boldsymbol{k} + \boldsymbol{G})^2} \left(1 - \mathrm{e}^{-((\boldsymbol{k}+\boldsymbol{G})^2 - \varepsilon)/\beta}\right), \tag{5}$$

not possessing any poles with respect to the energy since they are just compensated by the zeros of $\left[1 - \exp\left(\varepsilon - (\boldsymbol{k} + \boldsymbol{G})^2\right)/\beta\right]$. The badly converging sum (5) in $\boldsymbol{G}$-space may be transformed in a well converging sum in real space $\boldsymbol{R}$[1]):

$$A_{LL'}^{\boldsymbol{k}} = \sum_{\boldsymbol{R}} e^{i\boldsymbol{k}\boldsymbol{R}} A_{LL'}(\boldsymbol{R}). \tag{6}$$

The quantities $A_{LL'}(\boldsymbol{R})$ decrease very rapidly with increasing R since they contain integrals of type (for details, see Lehmann 83)

$$\int\limits_0^\infty \frac{\mathrm{d}k k^2 \mathrm{j}_l^2(kR_{\mathrm{MT}})\, \mathrm{j}_l(kR)}{(k^2 - \varepsilon)} \left(1 - e^{-(k^2-\varepsilon)/\beta}\right). \tag{7}$$

Terms like (5) may be included in (1) and (2) via matrix A. In evaluation of lattice sums, the split (4) and (5) is known already as the Ewald technique. The open parameter β has to be chosen in this way so that both an optimal decrease of (4) in $\boldsymbol{G}$-space and of (6) in $\boldsymbol{R}$-space is obtained.

2.6.2 The Basic Equations and Their Linearization

The quantities σ in (1) are diagonal with respect to L and L' and, besides terms $\mathrm{n}_l/\mathrm{j}_l$ from (2.5-9), also contain cotangents of the scattering phase shifts. Because of (2.3-8), the contribution for $l = 2$ shows the form

$$\sigma_2 = (\varepsilon_d - \varepsilon)/\Gamma_d + \tilde{\sigma}_2, \tag{8}$$

known from the TB method for atomic d-states. $\Gamma_d\tilde{\sigma}_2$ is a nearly energy independent expression which may be interpreted as a small correction of the resonance energy. The components of σ for $l = 0$ and 1 are of first order in $\cot \eta_l$. In the following they are denoted symbolically by σ_{11}. With the discrimination of s,p-parts (index 1) and d-parts (index 2), eqn. (2) may be written as

$$\det \begin{Vmatrix} \sigma_{11} & A_{12} & \Phi_1 \\ A_{21}^{+} & \sigma_2 + A_{22} & \Phi_2 \\ \Phi_1^{+} & \Phi_2^{+} & D - \varepsilon \end{Vmatrix} = 0. \tag{9}$$

[1]) The following identity for lattice sums has been used:

$$\sum_{\boldsymbol{G}} f(\boldsymbol{k} - \boldsymbol{G}) = \int \mathrm{d}^3k' f(\boldsymbol{k}') \sum_{\boldsymbol{G}} \delta(\boldsymbol{k}' - \boldsymbol{k} + \boldsymbol{G})$$

$$= \int \mathrm{d}^3k' f(\boldsymbol{k}')\, \Omega_0 (2\pi)^{-3} \sum_{\boldsymbol{R}} e^{i(\boldsymbol{k}-\boldsymbol{k}')\boldsymbol{R}} = \Omega_0 \sum_{\boldsymbol{R}} e^{i\boldsymbol{k}\boldsymbol{R}} \tilde{f}(\boldsymbol{R}),$$

where $\tilde{f}(\boldsymbol{r})$ represents the Fourier transform of $f(\boldsymbol{k})$:
$\tilde{f}(\boldsymbol{r}) = (2\pi)^{-3} \int \mathrm{d}^3k f(\boldsymbol{k})\, e^{-i\boldsymbol{k}\boldsymbol{r}}$.

Subtracting line 1 multiplied by $A_{21}^+\sigma_{11}^{-1}$ and $\Phi_1^+\sigma_{11}^{-1}$ from the second and third line, respectively, we obtain

$$\det \left\| \begin{matrix} \sigma_{11} & A_{12} & \Phi_1 \\ 0 & \sigma_2 + A_{22} - A_{21}^+\sigma_{11}^{-1}A_{12} & \Phi_2 - A_{21}^+\sigma_{11}^{-1}\Phi_1 \\ 0 & \Phi_2^+ - \Phi_1^+\sigma_{11}^{-1}A_{12} & D - \varepsilon - \Phi_1^+\sigma_{11}^{-1}\Phi_1 \end{matrix} \right\| = 0 \quad (10)$$

which represents a decoupling of s,p-states from the residual equations by taking them into account as a small (because of $\sigma_{11}^{-1} = \tan \eta'$) correction in the d-states and plane waves ($\boldsymbol{k} + \boldsymbol{G}$), respectively. In (10) the deciding disadvantages of the KKR equation are removed since (10) contains only in its diagonal strongly (but linearly!) energy dependent terms. The residual matrix elements $M(\varepsilon)$ depend only weakly on the energy and, therefore, may be approximated in the vicinity of an arbitrary energy (e.g. the physically interesting resonance energy ε_d) with good accuracy by the linear expression

$$M(\varepsilon) = M(\varepsilon_d) + M'(\varepsilon_d)\,(\varepsilon - \varepsilon_d)\,. \quad (11)$$

In this manner, we obtain from (10) the final representation of the basic equations

$$\det \|H - \varepsilon N\| = 0 \quad (12)$$

with the structure

$$H = \begin{pmatrix} H^{dd} & H^{ds} \\ H^{sd} & H^{ss} \end{pmatrix} \quad (13)$$

and

$$N = I - \left.\frac{\partial H}{\partial \varepsilon}\right|_{\varepsilon=\varepsilon_d}. \quad (14)$$

H^{dd} contains only the five d-states (dummy index: magnetic quantum number m). According to (8) and (10), it consists of the weakly changed atomic energies ε_d and of the "overlap integrals" $\Gamma_d A_{LL'}(\boldsymbol{R})$ from (6) and (7) to the nearest neighbours well known from the tight-binding method. Because of (8), matrices A have to be multiplied by the resonance width Γ_d. Therefore, Γ_d and the overlap of neighbouring wave functions determine the width of the d-bands. H^{ss} contains the energies $(\boldsymbol{k} + \boldsymbol{G})^2$ of free electrons moving in a pseudopotential arising from the weak scattering phase shifts $\tan \eta_l'$ (in the exact derivation contributions with $\tan \eta_2'$ from (8) have also to be considered). Thus, H^{ss} describes the nearly free electrons (NFE). The hybridization matrix elements $H^{sd} = H^{ds+}$ contain $\sqrt{\Gamma_d}\,\tilde{\Phi}_{L\boldsymbol{G}}$. They mix d-states and nearly free electron states in that region where both have nearly the same energy. Because of (4), the hybridization to the NFE states decreases with increasing energy. The matrix N in (14) consists of the unity matrix I and minor corrections $-\partial H/\partial\varepsilon$ which may be interpreted as orthogonalization corrections of the basis states.

N and H are Hermitian matrices. With this, the KKR equations (2.5-17) have been transformed rigorously into the *H-NFE-TB equations* (12) which may be solved by standard methods. The separation into tightly bound and nearly free states corresponds exactly to the physical situation in transition metals. Therefore, the H-NFE-TB equations are especially suited for the calculation of the band structure of such systems.

In order to simplify the derivation, we have so far assumed only one atom per unit cell. However, the H-NFE-TB equations are well suited for a generalization to several transition metal atoms (M_d) per unit cell. Then the rank of the matrix H equals the total number $5M_d$ of d-states in a unit cell. The number of NFE states to be taken into account depends on the desired accuracy. In the examples given in [3], their number amounts about seven to fifteen times the number of atoms per unit cell.

Moreover, losses of accuracy compared with the KKR equations may only be induced by the linearization (11). By appropriate choice of expansion energy ε_d in (11), these may be reduced as necessary.

The computational expense solving the H-NFE-TB equations which may be denoted as a linearized version of the KKR equations (LKKR) is about a factor 50 less than for the original KKR equations. Earlier versions were developed by Hubbard (69) and Pettifor (69, 72). An extended representation may be found in Lehmann (75 and 83).

2.6.3 Other Linearized Methods

Recently, linearized methods took an increasing importance for the treatment of complicated systems, because of their reduced computational expense. Also the linearized method of muffin-tin orbitals (LMTO) developed by Andersen (75) is based again on the KKR equations. However, the realization of that scheme does not depend on the MT potential. As experience shows that the results of band structure calculations in metals depend only weakly on the value of the constant potential between the MT spheres, it will be set equal to the energy sought. This has the extraordinary advantage that the structure constants for every lattice type have to be calculated only once, and then only for the energy value $\varepsilon = 0$. An extended description is given by Skriver (84). An especially efficient version is the so-called atomic sphere approximation (ASA) where the Wigner-Seitz cells are replaced by the insignificantly overlapping Wigner-Seitz spheres. Neglecting hybridization, canonical bands may be obtained which characterize a given lattice type. Then the band structure of a special metal results by continuous scaling of energy (according to the energy dependence of the scattering phase shifts).

Recently a TB version of the LMTO method has been developed (Andersen et al. 87), which makes possible the application to more complicated situations (alloys, extended defects, liquid and amorphous systems). It gives similarly as the optimized LCAO method (Eschrig 88) access to the full, non-spheridized electron density, needed for fast and accurate total-energy and force calculations.

The linearized APW approach (LAPW) developed by Andersen, Koelling and Arbmann (75) solves the Schroedinger equation inside the MT spheres not for the energy sought but for an energy $\varepsilon_l(\boldsymbol{k}+\boldsymbol{G})$, under the condition that at the MT radius not only does the wave function fulfil the continuity condition (2.4-1) to plane waves outside the MT spheres, but also its derivative. Therefore, the surface integral (2.4-5), inducing in (2.4-8) the inconvenient and strongly energy dependent sums over l, vanishes. Instead of that term, integrals of the wave functions at energies $\varepsilon_l(\boldsymbol{k}+\boldsymbol{G})$ over the MT spheres appear, which have to be multiplied by $\big(\varepsilon-\varepsilon_l(\boldsymbol{k}+\boldsymbol{G})\big)$ and, therefore, contain only a linear dependence on energy. A strong simplification results in the evaluation of the integrals by linearization of the energy dependence of the wave function $\big($here denoted with $R_l(\varepsilon, r)$ instead of $R_l(r, \varkappa)\big)$ around the energy ε_l:

$$R_l(\varepsilon, r) = R_l(\varepsilon_l, r) + (\varepsilon-\varepsilon_l)\,\dot{R}_l(\varepsilon_l, r) + \frac{(\varepsilon-\varepsilon_l)^2}{2}\,\ddot{R}_l(\varepsilon_l, r)\,. \quad (15)$$

The dots denote derivatives with respect to energy. By taking into account in (15) the second derivative, Smrcka (82) obtained a further improvement of accuracy in solving the secular equation (2.4-7) without an increase in computational expense. In addition to the linear KKR method, a quadratic version has also been developed (Faulkner and Beaulac 82).

2.7 Coherent Potential Approximation

Contrary to the approaches discussed so far, the *coherent potential approximation* (*CPA*) is appropriate for calculation of states in a disordered alloy. Here, the lattice sites are occupied randomly by different kinds of atoms. Frequently, this is called crystalline disorder as distinct from topological disorder in liquids or glasses.

The formulation of the CPA was first given by Soven (67). Its first application in the framework of tight binding was developed by Velicky, Kirkpatrick and Ehrenreich (68). For reasons of simplicity we follow this representation developed for an isolated band, even although for more realistic systems containing transition metal atoms, we require more complicated versions on the basis of the multiple scattering concept (KKR-CPA) corresponding to the KKR equations in ordered crystals. A good survey on the recent progress in this direction is given by Faulkner and Stocks (80), Faulkner (82), and Weinberger (90). A version on the basis of optimized LCAO (see [2.1.2]) yielding results with an accuracy similar to KKR-CPA but with considerably reduced numerical effort and improved physical transparency has been developed and tested by Richter, Eschrig, and Velicky (87). For the inclusion of relativistic effects (RKKR-CPA) see Weinberger and Gonis (87), Weinberger et al. (88), Weinberger (90).

In the one-band model the band index may be dropped. Then, the matrix elements of the Hamiltonian $H_{\boldsymbol{RR}'}$ are characterized by the lattice sites $\boldsymbol{R}$ and $\boldsymbol{R}'$ only. In analogy to (2.1-21) with $V_{\boldsymbol{RR}'} = V(\boldsymbol{R} - \boldsymbol{R}')$ and with $V(\boldsymbol{R})$ from (2.1-23'), we obtain

$$H_{\boldsymbol{RR}'} = \varepsilon^0_{\boldsymbol{R}} \delta_{\boldsymbol{RR}'} + H^0_{\boldsymbol{RR}'}; \quad H^0_{\boldsymbol{RR}'} = V_{\boldsymbol{RR}'}. \tag{1}$$

In the simplest case, we assume that in a binary alloy only the atomic levels $\varepsilon^0_{\boldsymbol{R}}$ fluctuate and that they assume the values ε^0_A and $\varepsilon^0_B = \varepsilon^0_A + \Delta$ depending on the occupation of the lattice site $\boldsymbol{R}$ by an atom type A or B, respectively. Furthermore, we assume that the overlapping integrals V to next nearest neighbours are the same as in the corresponding crystal ("diagonal" disorder). This means that the densities of states of the pure crystals A or B have the same form and are only shifted against each other by the value Δ.

A possible configuration of the alloy is given by the distribution of atoms of type A over the single lattice sites. For further conclusions we need the averages $\langle G \rangle$ of a quantity G over all configurations consistent with a given concentration c_A of atoms A. As quantity G, we consider in the following the Green function of the system defined in analogy to (2.5-3) (in (2.5-3) for free electrons)

$$\sum_{\boldsymbol{R}''} (H_{\boldsymbol{RR}''} - \delta_{\boldsymbol{RR}''}\varepsilon)\, G_{\boldsymbol{R}''\boldsymbol{R}'}(\varepsilon) = -\delta_{\boldsymbol{RR}'}. \tag{2}$$

G contains both information on eigenfunctions and eigenvalues of the Hamiltonian $H_{\boldsymbol{RR}'}$ (the poles of $G_{\boldsymbol{RR}'}(\varepsilon)$ just appear at the solutions of the homogeneous system of equations (2)). Via

$$D(\varepsilon) = -\frac{1}{\pi N} \operatorname{Im} \sum_{\boldsymbol{R}} G_{\boldsymbol{RR}}(\varepsilon) \tag{3}$$

and

$$n_{\boldsymbol{R}} = -\frac{1}{\pi} \int_{-\infty}^{\varepsilon_F} \mathrm{d}\varepsilon \operatorname{Im} G_{\boldsymbol{RR}}(\varepsilon) \tag{4}$$

both the density of states $D(\varepsilon)$ and the electron density $n_{\boldsymbol{R}}$ at lattice site $\boldsymbol{R}$ may be calculated from $G_{\boldsymbol{RR}'}(\varepsilon)$.[1]

We denote the configurational average of G by

$$\mathcal{G}_{\boldsymbol{RR}'}(\varepsilon) = \langle G_{\boldsymbol{RR}'}(\varepsilon) \rangle. \tag{5}$$

$\mathcal{G}_{\boldsymbol{RR}'}$ is translationally invariant (it depends only on the difference $\boldsymbol{R} - \boldsymbol{R}'$) as a consequence of averaging. $\mathcal{G}$ obeys an equation in analogy to (2) with a translationally invariant effective Hamiltonian $\mathcal{H}$ containing the so-called

[1]) The retarded Green function needed in (3) and (4) is obtained by adding a small imaginary part ($\varepsilon + i\delta$) to the energy ε in (2).

coherent potential $\Sigma(\varepsilon)$

$$\sum_{\boldsymbol{R}''} \left(\mathcal{H}_{\boldsymbol{R}\boldsymbol{R}''}(\varepsilon) - \delta_{\boldsymbol{R}\boldsymbol{R}''}\varepsilon\right) \mathcal{G}_{\boldsymbol{R}''\boldsymbol{R}'}(\varepsilon) = -\delta_{\boldsymbol{R}\boldsymbol{R}'}, \tag{6}$$

$$\mathcal{H}_{\boldsymbol{R}\boldsymbol{R}'} = H^0_{\boldsymbol{R}\boldsymbol{R}'} + \Sigma(\varepsilon)\,\delta_{\boldsymbol{R}\boldsymbol{R}'}. \tag{7}$$

Accordingly, $\mathcal{G}_{\boldsymbol{R}\boldsymbol{R}'}(\varepsilon)$ results simply by shifting the energy in the Green function $G^0_{\boldsymbol{R}\boldsymbol{R}'}(\varepsilon)$ belonging to an Hamiltonian H^0 in a periodic system by the first of all unknown coherent potential $\Sigma(\varepsilon)$:

$$\mathcal{G}_{\boldsymbol{R}\boldsymbol{R}'}(\varepsilon) = G^0_{\boldsymbol{R}\boldsymbol{R}'}\left(\varepsilon - \Sigma(\varepsilon)\right). \tag{8}$$

Consequently, $H_{\boldsymbol{R}\boldsymbol{R}'}$ and $\mathcal{H}_{\boldsymbol{R}\boldsymbol{R}'}$ differ by the fluctuating quantities

$$W_{\boldsymbol{R}\boldsymbol{R}'} = \delta_{\boldsymbol{R}\boldsymbol{R}'} W_{\boldsymbol{R}}; \quad W_{\boldsymbol{R}} = \left(\varepsilon^0_{\boldsymbol{R}} - \Sigma(\varepsilon)\right). \tag{9}$$

With (9) G may be expressed by $\mathcal{G}$,

$$G_{\boldsymbol{R}\boldsymbol{R}'} = \mathcal{G}_{\boldsymbol{R}\boldsymbol{R}'} + \sum_{\boldsymbol{R}''} \mathcal{G}_{\boldsymbol{R}\boldsymbol{R}''} W_{\boldsymbol{R}''} G_{\boldsymbol{R}''\boldsymbol{R}'}. \tag{10}$$

Iterating G on the righthand side, we obtain a representation by powers of the perturbation potential $W_{\boldsymbol{R}}$ which is suitable for configurational averaging, as the configuration is contained only in the quantities $W_{\boldsymbol{R}}$. (10) may be understood as the repeated scattering of an electron at perturbation $W_{\boldsymbol{R}}$, whose propagation from site $\boldsymbol{R}$ to site $\boldsymbol{R}'$ in the periodic crystal ($\mathcal{H}$) is described by $\mathcal{G}_{\boldsymbol{R}\boldsymbol{R}'}$.

Before performing the configurational average on the scattering centres independent of each other, repeated scattering at the same centre must be separated. This is best done by means of the so-called T-matrix

$$T_{\boldsymbol{R}\boldsymbol{R}'} = \sum_{\boldsymbol{R}''} W_{\boldsymbol{R}} G_{\boldsymbol{R}\boldsymbol{R}''} (\mathcal{G}^{-1})_{\boldsymbol{R}''\boldsymbol{R}'}, \tag{11}$$

which is obtained by comparing with (10),

$$T_{\boldsymbol{R}\boldsymbol{R}'} = \delta_{\boldsymbol{R}\boldsymbol{R}'} V_{\boldsymbol{R}} + V_{\boldsymbol{R}} \sum_{\boldsymbol{R}''} \mathcal{G}_{\boldsymbol{R}\boldsymbol{R}''} T_{\boldsymbol{R}''\boldsymbol{R}'}. \tag{12}$$

The separation of the terms $\boldsymbol{R}'' = \boldsymbol{R}$ from the sum in (12) gives

$$T_{\boldsymbol{R}\boldsymbol{R}'} = \delta_{\boldsymbol{R}\boldsymbol{R}'} t_{\boldsymbol{R}} + t_{\boldsymbol{R}} \sum_{\boldsymbol{R}''} \tilde{\mathcal{G}}_{\boldsymbol{R}\boldsymbol{R}''} T_{\boldsymbol{R}''\boldsymbol{R}'}, \tag{13}$$

where $\tilde{\mathcal{G}}$ agrees with $\mathcal{G}$ except $\tilde{\mathcal{G}}_{\boldsymbol{R}\boldsymbol{R}} = 0$ which excludes a repeated successive scattering at the same site. These scattering events are now contained in the t-matrix of the lattice site $\boldsymbol{R}$

$$t_{\boldsymbol{R}} = (1 - W_{\boldsymbol{R}} \mathcal{G}_{\boldsymbol{R}\boldsymbol{R}})^{-1}\, W_{\boldsymbol{R}}. \tag{14}$$

The definition (5) of $\mathcal{G}$ leads with (10) and (11) to the condition

$$\langle T_{\boldsymbol{R}\boldsymbol{R}'} \rangle = 0 \tag{15}$$

for $\Sigma(\varepsilon)$. The problem to be solved consists of averaging the product of the configuration dependent quantities t and T appearing in (13)

$$\langle tT\rangle = \langle t\rangle \langle T\rangle + \langle(t - \langle t\rangle)(T - \langle T\rangle)\rangle . \tag{16}$$

The neglec of the correlation described by the second term in (16) is termed single site approximation (SSA) since it takes into account via t exactly the scattering at a single site but disregards correlations induced by repeated scattering at site $\boldsymbol{R}$, but interrupted always by scattering processes at different sites $\boldsymbol{R}' \neq \boldsymbol{R}$. Thus, the SSA transmits (15) into the condition

$$\langle t\rangle = t_A c_A + t_B c_B = 0 \quad \text{with} \quad c_A + c_B = 1 , \tag{17}$$

where the concentrations c_A and c_B of both kinds of atoms are not independent.

Using (9) and (14) eqn. (17) may also be written as

$$\begin{aligned} &c_A\left(\varepsilon_A^0 - \Sigma(\varepsilon)\right) + c_B\left(\varepsilon_B^0 - \Sigma(\varepsilon)\right) \\ &-\left(\varepsilon_A^0 - \Sigma(\varepsilon)\right) G^0\left(\varepsilon - \Sigma(\varepsilon)\right) \left(\varepsilon_B^0 - \Sigma(\varepsilon)\right) = 0 . \end{aligned} \tag{18}$$

If the function $G^0(\varepsilon)$ is known, (18) represents an implicit equation for determination of the coherent potential $\Sigma(\varepsilon)$. According to (17) it has to be chosen such that in the effective medium at the single lattice sites, no scattering appears on average. Accurate analysis shows that the SSA does not differ from the exact solution below the fourth order in W. Moreover, one can show that the CPA may be understood as an expansion in powers of $1/Z$ where Z denotes the coordination number of the lattice. It means the CPA yields the best results for densely packed lattices.

The SSA considers in each case exactly the scattering at a lattice site in an effective medium. Therefore, it does not contain effects of short range order. Such effects in substitutionally disordered alloys are most successfully studied by means of the embedded-cluster method, see Gonis et al. (84), Erdös and Herndon (82).

3 Results of Band Structure Calculations

3.1 Simple Metals

Experimental results and theoretical calculations show universally that in the so-called simple (NFE) metals (fig. 3.1-1), where s- and p-electrons form the conduction bands, the band structure deviates only weakly from the parabolae (1.2-15) for free electrons. This means that the electrons move as described in [2.2] in a weak effective potential. The calculated band structure of aluminium (fig. 2.2-4) has already been described in this respect in [2.2.3]. Significant deviations from the free electron behaviour resulted only in the immediate vicinity of the Brillouin zone boundary where the minor discontinuity with a forbidden gap sketched schematically in fig. 1.2-3 appeared. According to the considerations of [4.1], this region corresponds exactly to the Bragg reflection. Therefore, the thence induced influence on the motion of the electron is determined exclusively by the geometry of the reciprocal lattice (location of the discontinuities).

Let us now discuss this uncomplicated, but for simple metals realistic, approximation in a transparent two-dimensional "model" metal varying the number of electrons per atom. In fig. 3.1-2 we study several Brillouin zones in the extended zone scheme. The circles correspond to Fermi surfaces for one to four electrons per atom. Since two electrons per atom just fill the area of the first Brillouin zone, the area of the circles corresponds to the n-fold of half a Brillouin zone. Figures 3.1-2b) to e) show the filling of the first to the fourth Brillouin zones in reduced and periodic zone schemes, respectively. In the case of one electron per atom, a Fermi "sphere" in the lowest conduction band is formed. For two electrons per atom, a part of the electrons flows into the second band (hatched regions) remaining holes in the first band (at the corners) which form the closed star-like figure A in the periodic zone scheme. In the second band, we obtain two lens-shaped regions B and C (again in the periodic zone scheme). For three electrons per atom (fig. d)), the hole region in the first zone is reduced further and the lens-shaped regions become more extended. A new situation arises for four electrons per atom (fig. e)). Then, the first band is filled completely. In the second band, we obtain the hole region D and the electrons missing here flow into the third and fourth band and form there regions E and F in the periodic zone scheme. The weak band gap causes a rounding off the corners of the Fermi surfaces.

In the three-dimensional case, the diversity of the shapes arising is much richer, depending on the symmetry of the lattice and the number of conduction

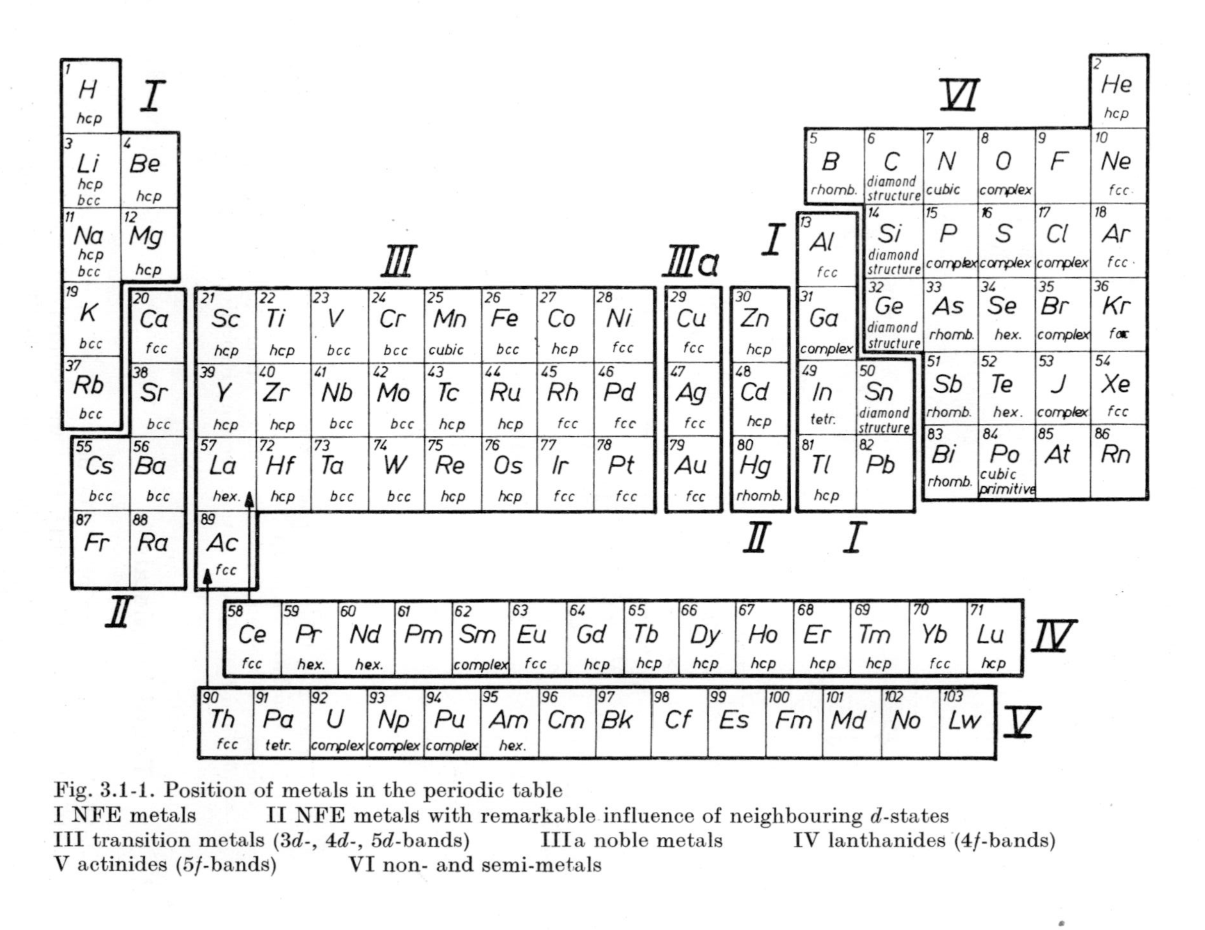

Fig. 3.1-1. Position of metals in the periodic table
I NFE metals II NFE metals with remarkable influence of neighbouring d-states
III transition metals ($3d$-, $4d$-, $5d$-bands) IIIa noble metals IV lanthanides ($4f$-bands)
V actinides ($5f$-bands) VI non- and semi-metals

electrons per atom. In the single bands, the Fermi surfaces are composed of pieces cut away from the Fermi sphere by planes of discontinuities (1.2-22). For the univalent alkaline metals, the unperturbed Fermi sphere does not yet touch the Brillouin zone boundary. In this case, the real Fermi body generated by the influence of the lattice potential shows smaller deviations from ideal sphericity than manufactured billiard-balls!

The band structures for all NFE metals are available with high accuracy (mainly pseudopotential and APW calculations, respectively).

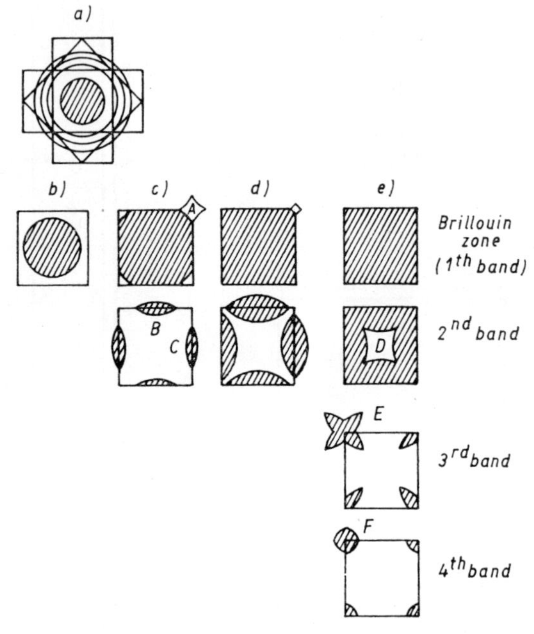

Fig. 3.1-2. a) BZ in the extended zone scheme for a two-dimensional model, Fermi surfaces for nearly free electrons
b) to e) representation in the reduced scheme for one (b)), two (c)), three (d)), and four (e)) conduction electrons, respectively

3.2 Transition Metals

The stable low-temperature phases of pure transition and noble metals (see fig. 3.1-1) will be discussed in [5.3.1]. Here we mention only a strict correlation between structure and the number of conduction electrons per atom N_c.

Transition and noble metals show band structures very similar to each other characterized by a narrow resonance of d-electrons with width Γ_d at energy ε_d near the atomic d-levels. Its consequences for the band structure have already been discussed in [2.3.3]. The interaction of the free and tightly bound electrons causes the hybridization presented schematically in fig. 3.2-1, which is characteristic for every transition and noble metal.

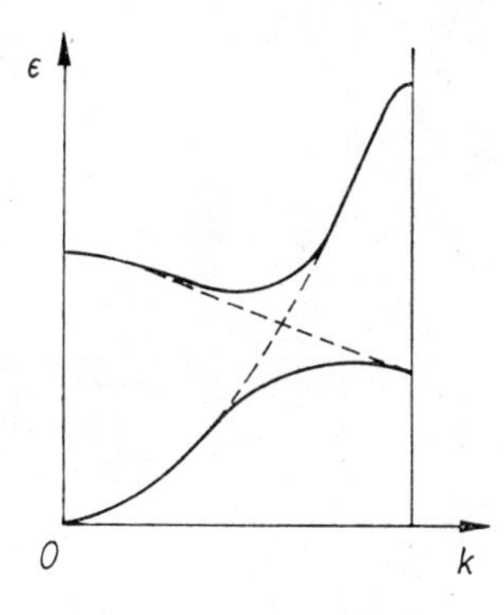

Fig. 3.2-1. Hybridization of the free electron band and the d-band (schematically for transition and noble metals). The dashed lines refer to the non-hybridized bands

Examples of band structure and density of states for metals with *fcc* lattices (Cu) and *bcc* lattices (V), respectively, are given in figs. 1.2-7 and 3.2.-2. As a consequence of the dominating meaning of d-states, the band structures of different transition and noble metals with the same lattice type differ only by the position ε_d and width Γ_d of the d-band and by the position of the Fermi energy ε_F which is determined by the number of conduction electrons per atom. Noble metals possess five filled d-bands (occupied with 10 electrons altogether), meaning that ε_F lies above the d-band in a band of nearly free electrons. Therefore, the Fermi surface has a nearly spherical shape (compare fig. 1.2-5) as in alkaline metals. The most significant deviations from their shape arise near the point L on the Brillouin zone boundary where the Fermi body exhibits necks to the neighbouring zone. The reason for this lies in the fact that the energy of the point L_2 is situated below ε_F and that therefore the states on the line $\overline{L_2U}$ in the direction of U are occupied up to the point marked in fig. 1.2-7 by an arrow.

Inside a given period of transition metals, Γ_d decreases with increasing atomic number as the resonance energy is lowered with respect to the MT potential in the interstitial region and, therefore, the resonance becomes narrower. In figs. 1.2-7 and 3.2.-2, the schematic course of the bands in fig. 3.2-1 may be seen clearly. Starting from point $\Gamma(\boldsymbol{k} = 0)$, we observe the parabola-like bands of the nearly free electrons; their behaviour is interrupted by the five narrow bands of d-electrons. In most cases, they are degenerated on the marked lines with high symmetry. This is not the case for the direction $\overline{\Gamma N}$ in fig. 3.2-2.

The density of states for copper in fig. 1.2-7b) shows a structure with three peaks typical for *fcc* metals. Near the Fermi energy, the density of states has already dropped considerably. The hybridization on the line $\overline{\Gamma X}$ causes an

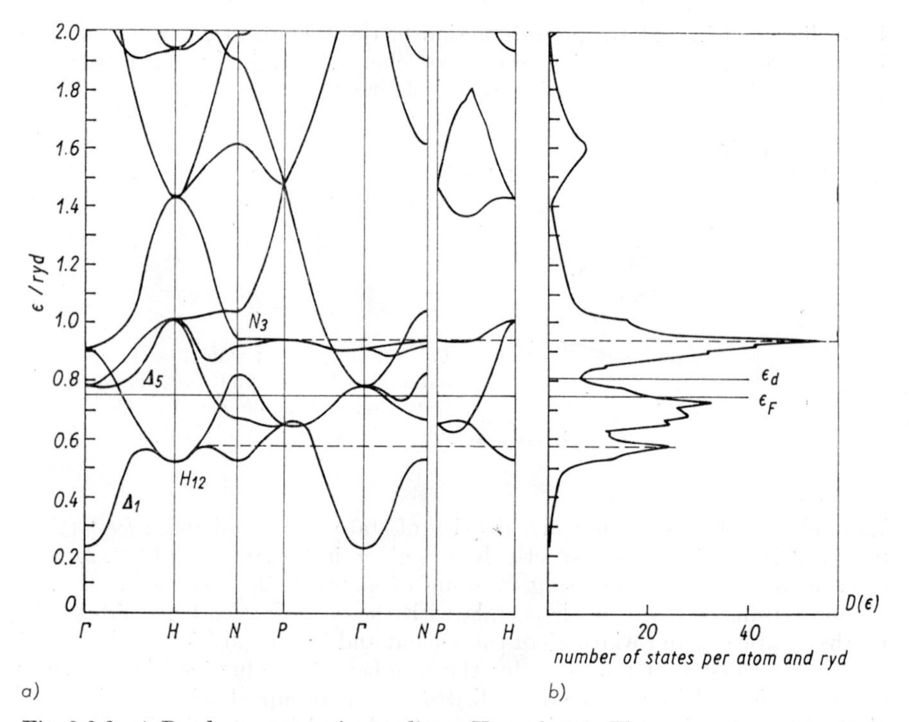

Fig. 3.2-2. a) Band structure of vanadium (Urwank 74). The symmetry points Γ, H, N, P of the *bcc* structure are denoted according to the description in fig. 1.2-5
b) DOS of vanadium. The strong spikes in $D(\varepsilon)$ are caused by the marked regions in fig. a) where the bands are very flat and, therefore, the states are very dense. These energies may be observed easily (see [4.3.])

energy gap between the first and second band which manifests itself in the density of states as a weak saddle at the beginning of the d-band.

The band structures of other *fcc* transition metals Rh, Pd, Ir, and Pt show a great similarity to those of the noble metals. As a consequence of the unfilled d-bands, the Fermi surfaces take on a more complicated shape and, as a rule, consist of several parts.

As an example of *bcc* metals, in fig. 3.2-2 the band structure of vanadium is given. The points H_{12} and N_3 in fig. 3.2-2 represent pure unhybridized d-states and, therefore, fix the upper and lower limits of the d-band. The deep minimum of DOS near 0.6 *ryd* is characteristic for *bcc* metals, and plays an important role in the stability of the *bcc* structure. In the direction $\overline{\Gamma H}$ we observe in the Δ_1-band a strong hybridization, whereas the Δ_5-band interacts only weakly with NFE states of higher energy.

Most of the band structure calculations for this group have been done for niobium (Mattheiss 70), and considerable experimental data are also available (for a compilation, see Mackintosh and Andersen 80, for example).

Because of its antiferromagnetic ordering at low temperatures in the form of static spin density waves along the cubic axis [100], chromium excites much interest. The large nearly plane region of the Fermi surface in the paramagnetic state is responsible for this peculiarity.

Since the *hcp* metals have two atoms per elementary cell, they show a more complicated band structure. The essential difference between both groups with $N_c = 3, 4$ and $N_c = 7, 8$ lies in the fact that ε_d and Γ_d decrease with increasing N_c. A detailed discussion of the *hcp* transition metals is given by Jepsen et al. (82).

The one-particle model and resulting band structures also prove to be quite successful for understanding other physical properties of transition metals. If the band width is of the order of the exchange interaction, many-body effects become important (described for example by the Hubbard model). This is only the case in alloys (e.g. oxides) of transition metals, if because of the larger distances between metallic atoms the band widths decrease.

The trends of the important potential parameters which in densely packed metallic structures are nearly independent of the crystalline symmetry may essentially be explained on the basis of atomic properties. However, especially for heavier metals (4*d*, 5*d*), the inclusion of relativistic effects is necessary (Nemoshkalenko et al. 82), for the electronic structure and electronic properties of actinide systems see Weinberger and Gonis (87). For a survey on band structures of transition metals, see Mackintosh and Andersen (80).

The magnetic moments in the ground state (temperature zero) of spin-polarized first principles calculations for Fe, Co, and Ni without using any fitting parameters agree very well with the experimental data: 2.15 (Fe); 1.56 (Co); 0.59 (Ni) in units of μ_B per atom ($\mu_B = |e|\, \hbar/2m$), see Moruzzi et al. (78). Self-consistent spin-polarized band structure calculations for antiferromagnetic chromium have been performed by Kuebler (80) and Skriver (81). The onset and approach to the Hund's rule limit in *bcc* 3*d* transition metals is studied by Moruzzi and Marcus (88).

The magnetic structure of ordered compounds has also been calculated on that basis.

The formation of magnetic moments in disordered NiCu and FeCo has recently been treated by Richter and Eschrig (87) on the basis of a spin-polarized calculation using the first principle LCAO-CPA (Richter et al. 87).

The quoted papers on ferromagnetically or antiferromagnetically ordered metals concern only the ground state, i.e. properties at zero temperatures. Previous attempts of a theory for non-zero temperatures were based either on the assumption that the local magnetic moments are essentially unchanged at higher temperatures (Heisenberg model, well founded for insulators) and only undergo changes in direction (spin fluctuations), or on the assumption that the orientation of the local moments remains unchanged and the absolute value reduces with increasing temperature (Stoner, band, or itinerant model:

for a review, see Herring 66). Recently, the electronic theory of magnetically ordered metals makes important contributions to the understanding of magnetic properties of metals near the Curie temperature T_c of the phase transition from ferromagnetic to paramagnetic state combining aspects of both theories. Such a theory for finite temperatures has to give an understanding and with it the possibility of evaluation of the following properties from first principles: the Curie temperature T_c, the temperature dependence of the local magnetic moments (which also exist above T_c), the magnetization $M(T, H)$ for $T > T_c$, the Curie-Weiss magnetic susceptibility and the observed magnetic short range order above T_c, and the spectrum of the elementary excitations of magnetically ordered systems (spin waves). There currently exist two controversial main attempts. At one extreme are the disordered local moment (DLM) theories which assume no correlation among the different lattice sites. The most sophisticated calculation was done by Gyorffy et al. (84) on the basis of multiple scattering theory (KKR-CPA). At the other extreme are the local band theories (LBT) which assume an important degree of short range order in the paramagnetic state.

Fulde (84) reports a simple but effective method by including many-particle corrections into the band model to describe local magnetic moments above T_c and to solve the discrepancies mentioned (see also Oles and Stollhoff 84).

The starting theory of surface magnetism (Mukherjee et al. 82) concerns, for example, the explanation of the magnetic properties of thin Fe layers on Ag single crystals (Gay and Richter 86), of Mn layers (Oguchi and Freeman 86), and of Ni layers (Moruzzi et al. 86).

3.3 Compounds and Alloys

In intermetallic compounds, the various kinds of atoms may occupy the lattice sites both in ordered form and thus initiate superlattices, but may also be distributed randomly as in mixed crystals. The latter case of disordered systems requires methods such as CPA treated in [2.7].

In alloys, the same principles hold as in pure metals, but several peculiarities should be taken into account. The biggest problems arise in calculating the crystal potential. Because of the stronger local fluctuations of the charge density in alloys and the reduced symmetry of a given atom with respect to pure metals, the evaluation of the potential is more difficult than in pure metals. Charge transfer effects play a decisive role and self-consistent calculations are needed.

With the increasing number of atoms per elementary cell, the numerical effort increases so that computer time and costs limit the calculations. Therefore, in APW and KKR calculations for complicated structures, the energy spectrum is evaluated only at a few $\boldsymbol{k}$-points with high symmetry where the eigenvalue problem may be simplified considerably for symmetry reasons.

Then, the intermediate regions are described by interpolation methods such as the TB approach with fitting parameters. In this way, a dense network of $\boldsymbol{k}$-points may be obtained which is necessary for the calculation of DOS and further measurable quantities. Using this approach, one should be aware of larger uncertainties especially concerning the wave functions. Because of the reduced numerical expense in these cases, the application of the H-NFE-TB method or other linearized versions and of the optimized LCAO method treated in [2.1] is of large advantage.

As a consequence of the large number of atoms and electrons per elementary cell, on one hand the lattice constant a increases and the extension of the Brillouin zone $\sim a^{-1}$ decreases, and on the other hand essentially more occupied bands appear than in pure metals. Therefore, the DOS is more structured. As an example the DOS of the Laves phase $MgZn_2$ with 12 atoms per elementary cell is given in fig. 3.3-1. Although the pseudopotential is weak, it represents a relatively strong perturbation for the many densely lying bands of free electrons. As a consequence, the behaviour of nearly free electrons is not observed easily in the band structure. The DOS in fig. 3.3-1 fluctuates around the dashed straight line for free electrons. An overview on the results obtained for $MgZn_2$ is given by Ziesche (77).

As a further example, we treat the (non-magnetic) intermetallic compound FeAl. This crystallizes in a *bcc* structure denoted as B2 or CsCl structure. The atoms of one kind are situated at the midpoint of a cube and those of the other kind at its corners. In a certain sense, this structure serves as the simplest model of an ordered alloy exhibiting the essential characteristics of intermetallic compounds.

The band structure calculated by Ziesche et al. (78) by means of the H-NFE-TB method is given in fig. 3.3-2. The parabola-like NFE band around

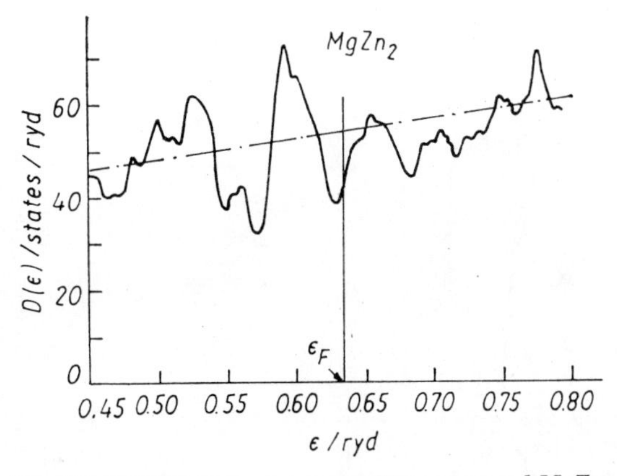

Fig. 3.3-1. DOS of the intermetallic compound $MgZn_2$ (Lehmann et al. 70). It shows a marked distinctly structure. The DOS of free electrons with a corresponding spatial density is denoted by the dash-dot line

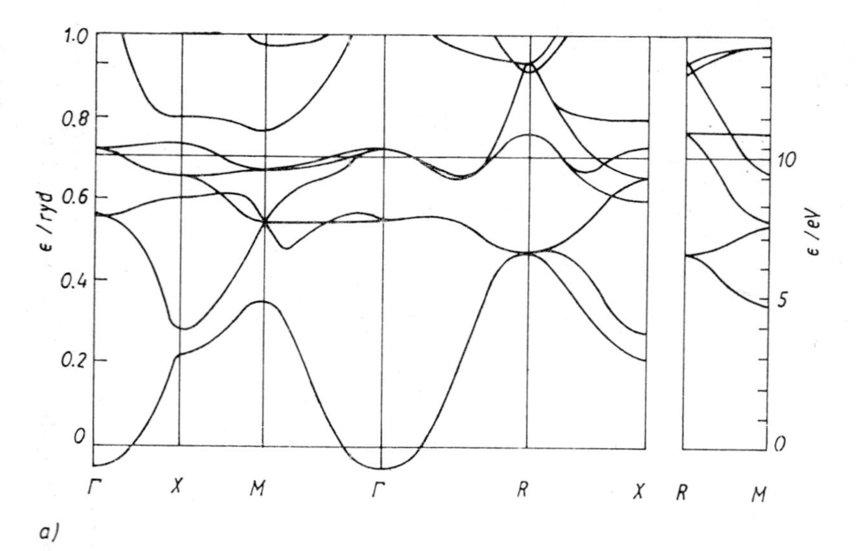

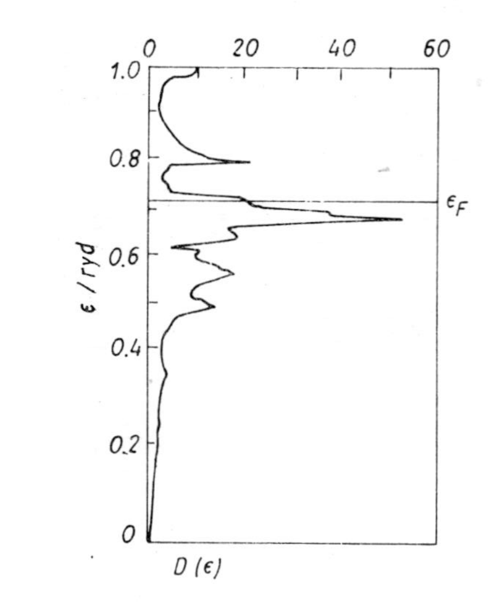

Fig. 3.3-2. a) Band structure of FeAl (B2 structure), after Ziesche et al. (78), lattice constant $a = 5.4973 a_0$
b) DOS of FeAl (in units: states per *ryd*, elementary cell, and spin). The notation of the points X and R with increasing energy is: $X_{4'}$, X_1, X_2, X_5, X_3, X_1 and $R_{25'}$, R_{12}, $R_{12'}$, R_{15}, respectively

point Γ can be seen clearly. Because of the smaller size of the BZ contrary to pure metals, the weak energy gap is situated below the d-band (for point X at levels $X_{4'}$ and X_1). At point R, the gap in the NFE band overlaps with the region of d-bands. The width of the d-bands is given by the separation of levels $R_{25'}$, and X_1. In the series VAl, CrAl, FeAl, CoAl, NiAl (Mueller et al. 78) it decreases from 450 to 235 mryd. This trend has the same cause as in pure transition metals. In the alloy with Al, the width of the d-bands reduces in comparison with the pure transition metal, though the resonance width is somewhat smaller in the pure metal than in the alloy. This behaviour, surprising at first glance, shows that the band width depends both on the property of a single scattering centre and on distance, number, and kind of neighbours in the crystal. In the series from VAl to NiAl the Fermi energy increases because of the increasing number of conduction electrons. Its position in VAl is near the low-energy flank of the strong maximum of DOS, in CrAl at its top, in FeAl at its high-energy flank near 710 mryd, in CoAl in the region of the following deep minimum, and in NiAl above the next maximum outside the d-band. The stability of a system is large if the Fermi energy is located at the upper boundary of a region with high DOS. This condition is fulfilled best in CoAl, but also in FeAl and NiAl. This result corresponds to the observed transition temperature of the homogeneous B2 phase (FeAl: 1250 °C, CoAl: 1645 °C, NiAl: 1638 °C). Compared to this, the B2 phase of CrAl is not stable. The H-NFE-TB band structure calculations agree well with the observed X-ray fine-structures.

The electronic structure of further intermetallic compounds is treated by Rennert and Lehmann (77), for example.

The peculiarities of disordered alloys will be discussed only briefly using the schematic behaviour of the DOS (fig. 3.3-3). For the ordered system (thick line), it shows two bands separated by a forbidden gap with marked maxima and minima and sharp band edges with the expected parabola-like singularities. In the disordered system (dashed line) in the forbidden region, defect levels ① may appear which at higher defect concentrations form defect bands. The sharp band edges are smeared out to band tails ② as a consequence of statistical fluctuations of the atomic distribution. In the cases ① and ②, perturbation

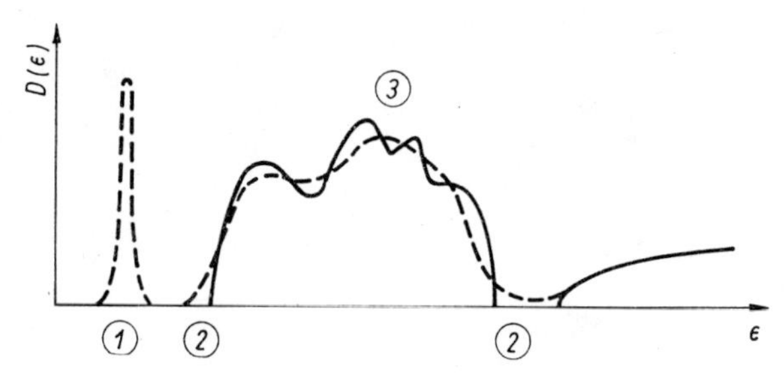

Fig. 3.3-3. DOS in disordered alloys

theory is invalid because in the relevant region in the unperturbed system no states exist at all. These regions are of special interest in semiconductors and insulators as the Fermi energy is situated there. In metals, band region ③ is important. There we observe in the disordered system a smoothing of spikes of DOS, since energy levels are broadened because of the finite life time of momentum states, as a consequence of the loss of periodicity. For not too strong perturbation, the wave functions in the regions ③ are still extended whereas the states in regions ① and ② are spatially localized.

A survey on band structure calculations published up to 1970 is given by Dimmock (71). A recent reviev is given by Weinberger (90). — Some compounds and alloys are of direct technological interest. Examples are refractory compounds, i.e. transition metal nitrides and carbides. They are used as cutting tools (TiN) or superconductors ($Nb\,C_{0.3}\,N_{0.7}$). Hardness and T_c are influenced by vacancies, which are contained in these materials normally. Therefore, an understanding of vacancy-induced changes in the electronic structure is important (Redinger et al. 87). — Compounds and alloys of rare earths with Fe, Co, Cu, Zr (and B) are of technological interest as new permanent magnets.

4 Experimental Determination of Band Structure

4.1 Dynamics of Lattice Electrons

So far, we have investigated only the stationary eigenstates of electrons in a periodic potential. Non-stationary processes are of importance under the influence of external fields.

For the treatment of the problems thereby arising, we represent an electron by a wave packet relatively narrow in $\boldsymbol{k}$-space but extended over many lattice cells in real space. Assuming that the wave packet consists of wave functions of only one band, no band transitions should be described, and the function $\varepsilon(\boldsymbol{k})$ is unequivocal and periodic.[1])

These assumptions are fulfilled in the most interesting cases. Therefore, it is sufficient to describe the motion of the centre of gravity of the wave packet. An example where interband transitions are of importance is the so-called Zener breakdown ("inner" field emission) in insulators and semiconductors. In this case an electronic transition from a fully occupied band, to an empty one across a forbidden energy gap Δ is induced by a strong electric field $\boldsymbol{E}$ with

$$e\,|\boldsymbol{E}|\,a \gtrapprox \Delta^2 \bigg/ \left(\frac{\hbar^2}{2m}\,\frac{\pi^2}{a^2}\right). \tag{1}$$

The electron in the empty band is accelerated by the external field and thus causes an electric current. $e\,|\boldsymbol{E}|\,a$ is the energy gained at the distance of a lattice constant a under the action of the external field. $(\hbar\pi/a)^2/2m$ corresponds to the band width for free electrons.

In the one-band approximation, because of the periodicity of the energy function, the momentum operator $\hat{\boldsymbol{p}}$ in (2) can be replaced by $(\hbar/i)\,\partial/\partial\boldsymbol{r}$ as in the case of free particles. Therefore, the Hamiltonian can be chosen as[2])

$$H = \varepsilon_\nu(\hat{\boldsymbol{p}}/\hbar) = \tilde{\varepsilon}_\nu(\hat{\boldsymbol{p}})\,. \tag{2}$$

[1]) If the crystal properties change over regions large in comparison with the extension of a wave packet, then a weak spatial dependence of the dispersion relation $\varepsilon(\boldsymbol{k}, \boldsymbol{r})$ may be assumed. This approach is used often in semiconductor physics (near interfaces). It will fail if the spatial variation of ε is of the order of the lattice parameters.

[2]) In this chapter, operators such as $\hat{\boldsymbol{r}}$ and $\hat{\boldsymbol{p}}$ are denoted by a sign $\wedge$. Averages of these operators over wave packets are denoted by a bar: $\overline{\boldsymbol{r}}$, $\overline{\boldsymbol{p}}$.

Then, the Hamiltonians (2) and (1.2-1) have the same matrix elements with respect to functions of band ν.[1]) The velocity $\boldsymbol{v}$ of the electron equals the group velocity of the wave packet:

$$\boldsymbol{v} = \frac{\partial \omega}{\partial \boldsymbol{k}} = \frac{1}{\hbar} \frac{\partial \varepsilon}{\partial \boldsymbol{k}}. \tag{3}$$

The same relation may also be obtained (with position operator $\hat{\boldsymbol{r}} = -\frac{\hbar}{i}\frac{\partial}{\partial \boldsymbol{p}} = i\frac{\partial}{\partial \boldsymbol{k}}$) via the Ehrenfest theorems

$$\overline{\boldsymbol{v}} = \dot{\overline{\boldsymbol{r}}} \equiv \frac{i}{\hbar}\overline{[H, \hat{\boldsymbol{r}}]} = \frac{i}{\hbar}\overline{\left[\tilde{\varepsilon}(\boldsymbol{p}), -\frac{\hbar}{i}\frac{\partial}{\partial \boldsymbol{p}}\right]} = \frac{1}{\hbar}\frac{\partial \varepsilon}{\partial \boldsymbol{k}}. \tag{3'}$$

For external forces $\boldsymbol{F}$ ($\boldsymbol{F}$ constant) we obtain

$$H' = \tilde{\varepsilon}(\hat{\boldsymbol{p}}) - \boldsymbol{F}\boldsymbol{r}; \quad \boldsymbol{F} = -\frac{\partial V}{\partial \boldsymbol{r}}. \tag{4}$$

The acceleration $\ddot{\overline{\boldsymbol{r}}}$ of the centre of gravity may be obtained again by the Ehrenfest theorem[2]):

$$\ddot{\overline{\boldsymbol{r}}} = \frac{i}{\hbar}\overline{\left[\varepsilon(\boldsymbol{k}) + \frac{1}{i}\boldsymbol{F}\frac{\partial}{\partial \boldsymbol{k}}, \quad \frac{1}{\hbar}\frac{\partial \varepsilon}{\partial \boldsymbol{k}}\right]} = \boldsymbol{F}\,\overline{\frac{1}{\hbar^2}\frac{\partial}{\partial \boldsymbol{k}} \circ \frac{\partial}{\partial \boldsymbol{k}}\,\varepsilon}. \tag{5}$$

Analogously, for the time-derivative of the momentum

$$\dot{\overline{\boldsymbol{p}}} = \frac{i}{\hbar}\overline{\left[\varepsilon(\boldsymbol{k}) + \frac{1}{i}\boldsymbol{F}\frac{\partial}{\partial \boldsymbol{k}}, \quad \hbar\boldsymbol{k}\right]} = \boldsymbol{F} \tag{6}$$

we obtain an expression similar to the classical result. Comparing (5) with the Newton equation of motion, $\ddot{\boldsymbol{r}} = \boldsymbol{F}/m$ shows that $1/m$ has to be replaced by the (symmetric, momentum dependent) reciprocal mass tensor $1/\mathbf{M}$,

$$\frac{1}{\mathbf{M}} = \frac{1}{\hbar^2}\frac{\partial}{\partial \boldsymbol{k}} \circ \frac{\partial}{\partial \boldsymbol{k}}\,\varepsilon. \tag{7}$$

[1]) Introducing the Hamiltonian H in (2) in the one-band approximation, it is especially surprising that with $\varepsilon_\nu(\hat{\boldsymbol{p}}/\hbar)\,\psi_{\boldsymbol{k}} = \varepsilon_\nu(\boldsymbol{k})\,\psi_{\boldsymbol{k}}$ the wave vector $\boldsymbol{k}$ is represented formally as the eigenvalue of the true momentum operator $\hat{\boldsymbol{p}}$, although the Bloch functions are certainly not eigenfunctions of $\hat{\boldsymbol{p}}$. In reality, it is not $\hat{\boldsymbol{p}}$ which acts on the Bloch functions, but a function $\varepsilon_\nu(\hat{\boldsymbol{p}}/\hbar)$ periodic in the reciprocal lattice because of (1.2-12). Let us consider only the component $\exp(i\boldsymbol{k}\boldsymbol{R})$ in the Fourier representation of $\varepsilon_\nu(\boldsymbol{k})$. Then the operator $\exp(i\hat{\boldsymbol{p}}\boldsymbol{R}/\hbar)$ simply means the translation operator T in real space which generates a translation by the lattice vector $\boldsymbol{R}$. Regarding T, the Bloch functions are eigenfunctions with the eigenvalue $\exp(i\boldsymbol{k}\boldsymbol{R})$. Thus the Bloch functions also represent eigenfunctions of the Hamiltonian (2) with eigenvalues $\varepsilon_\nu(\boldsymbol{k})$.

[2]) The circle ∘ in (5) denotes the dyadic product.

For illustration, we will discuss the one-dimensional motion of an electron in the lowest band under the influence of a constant external force in the approximation of nearly free electrons. The perturbation of the energy band of free electrons is remarkable only near $k = G/2$. In fig. 4.1-1 the quantities ε, v, and $1/M$ are shown as functions of k. The electron with initial velocity $v_0 = 0$ is accelerated like a free one until k approaches the zone boundary $G/2$. Then, in spite of the constant force, the velocity reduces again to zero (zone boundary), and afterwards becomes negative. This means, the real space wave packet runs opposite to the external force and executes an oscillatory motion running a distance $\hbar G$ in momentum space. (This statement is not completely correct, since with a certain probability the particle may do a Zener break-through into a band with higher energy.) At the zone boundary the electron behaves like a "donkey-particle": it is tugged one way, but moves exactly in the opposite direction. However, this is simply a different description of the phenomenon of Bragg reflection, appearing if the reflection condition (1.2-22) or condition $\lambda = 2a$ for the wave length λ of the electron is fulfilled. With the discussion of fig. 4.1-1, the different conductivity behaviour of substances (electrons, holes) may also be explained.

In the presence of magnetic fields $\boldsymbol{B}$ which are not too strong (for the remaining concentrated wave packets still to be sensible, see also [4.2]), the above equations of motion still hold, if the canonical momentum $\boldsymbol{p}$ is replaced by the mechanical one $\boldsymbol{q} = \boldsymbol{p} - e\boldsymbol{A}$ (the average bars for the concentrated wave

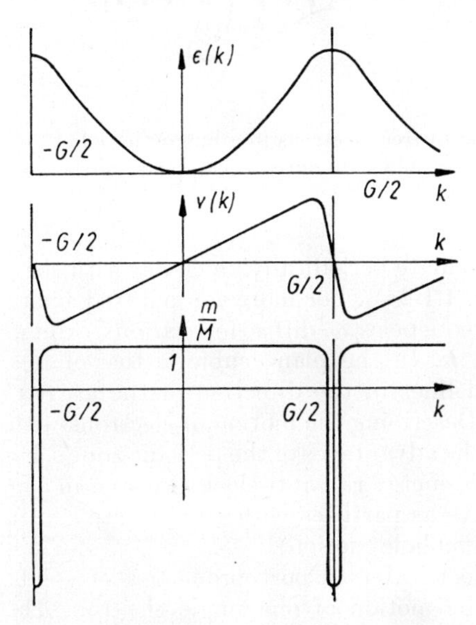

Fig. 4.1-1. Energy ε, velocity v, and effective mass m/M of a lattice electron as a function of wave number k

packets are omitted in the following, $\boldsymbol{A}$ denotes the vector potential belonging to the magnetic field $\boldsymbol{B}$):

$$\dot{\boldsymbol{r}} = \frac{1}{\mathbf{M}} \dot{\boldsymbol{q}}; \quad \dot{\boldsymbol{q}} = e\dot{\boldsymbol{r}} \times \boldsymbol{B}; \quad \boldsymbol{B} = \frac{\partial}{\partial \boldsymbol{r}} \times \boldsymbol{A}. \tag{8}$$

In this semiclassical approximation after multiplication by $\dot{\boldsymbol{r}}$, (8) may be integrated to the law of energy conservation:

$$\frac{\mathrm{d}\bar{\varepsilon}(\boldsymbol{q})}{\mathrm{d}t} = \dot{\boldsymbol{q}} \frac{\partial \bar{\varepsilon}(\boldsymbol{q})}{\partial \boldsymbol{q}} = \dot{\boldsymbol{q}}\dot{\boldsymbol{r}} = 0 \curvearrowright \bar{\varepsilon}(\boldsymbol{q}) = \text{const}. \tag{9}$$

In the same way, one obtains after multiplication of (8) by $\boldsymbol{B}$

$$\boldsymbol{q}\boldsymbol{B} = \text{const.} \tag{10}$$

Therefore, the motion occurs as in the classical case in $\boldsymbol{q}$-space on the intersection line of the surface of constant energy with the plane normal to the direction of magnetic field $\boldsymbol{B}$.

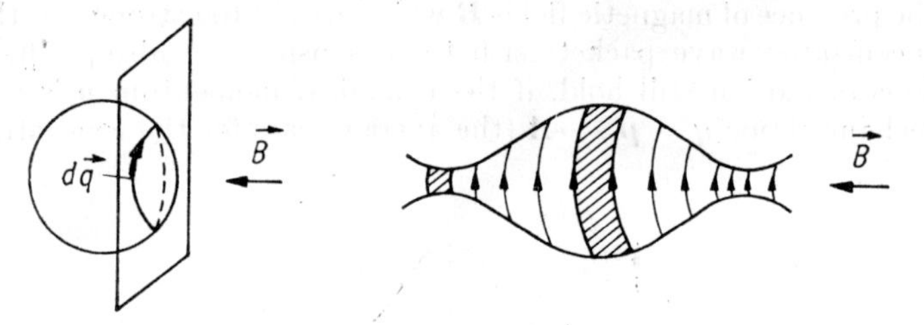

Fig. 4.1-2. a) Motion of free electrons in a magnetic field
b) Motion of electrons on a cylinder-shaped energy surface in a magnetic field

Therefore, free electrons circulate periodically on circles with the same angular velocity $\omega = |e|\, B/m$ (fig. 4.1-2a)). For more complicated energy surfaces (perhaps fig. 4.1-2b)), the frequencies ω differ for various orbits, since $1/\mathbf{M}$ according to (7) depends on $\boldsymbol{k}$. In the plane cubic lattice of fig. 4.1-3, the electrons run on the marked lines in the direction of the arrows (if $\boldsymbol{B}$ goes into the plane of the paper). Describing the motion of electrons with an energy ε_5 (near the top of the band), the advantages of the periodic zone scheme become evident again. Particles of this energy run anti-clockwise, and in comparison to ordinary electrons they behave as particles with an opposite e/m ratio (which is essential for interpretation as hole motion).

For the explanation of electrical transport properties of solids, it is not sufficient to consider only the motion of one single electron. The ensemble of all electrons as a whole participates in this macroscopic effect. Switching on a constant electrical field $\boldsymbol{E}$, the Fermi sphere (for the present we consider a nearly constant crystal potential) according to fig. 4.1-4 is shifted during the time element δt in momentum space by a quantity $\delta \boldsymbol{p} = e\boldsymbol{E}\delta t$ since every momentum state obtains an additional momentum $\delta \boldsymbol{p}$ with respect to (6). Thus, the system moves into a state of higher energy than the ground state. The total momentum $\boldsymbol{P}$ no longer equals zero, i.e. a mass current and concomitant electrical current flows. In reality, this distribution of states occupied

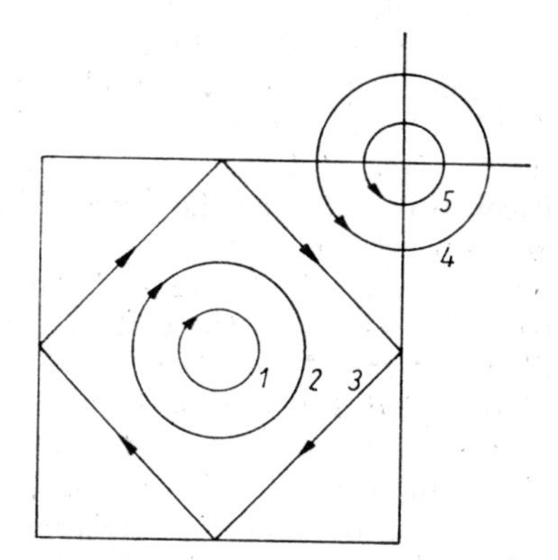

Fig. 4.1-3. Circulation direction of electrons on surfaces of constant energy in a magnetic field

by electrons will be disturbed by defect atoms and lattice oscillations, respectively (in the presence of such perturbations the total momentum is no longer a conserving quantity), and thus the scattering of electrons into unoccupied states with lower energy $1 \rightarrow 2$ begins. In this way the original state with $\boldsymbol{P} = 0$ will be restored. Thus, the effect of an applied external short-time field $\boldsymbol{E}$ vanishes after the characteristic relaxation time τ. Only for fields

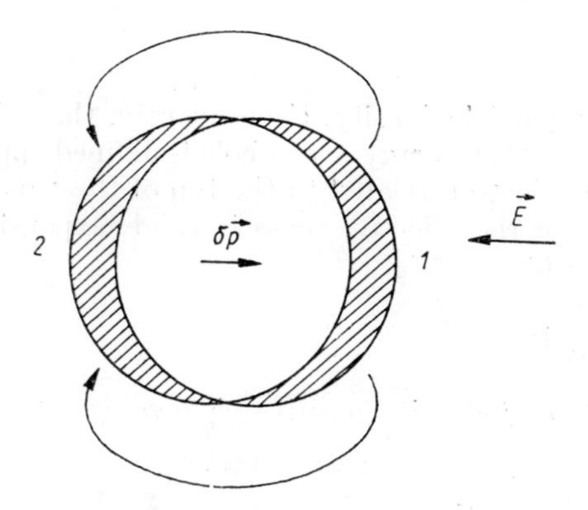

Fig. 4.1-4. Interpretation of conductivity

applied continuously does a stationary state result. In this state in first approximation the Fermi sphere is shifted by a small momentum $\delta \boldsymbol{p}$ proportional to $\boldsymbol{E}$. It is remarkable that only a few states (hatched in fig. 4.1-4) participate in scattering and transport processes. Therefore, considerable interest in detailed knowledge in the structure of the Fermi surface arises.

In superconductors, the mentioned scattering of momentum states is prevented by an "energy gap" of width Δ immediately at the Fermi surface, and therefore, an undamped current may flow after switching off the external electrical field. As we will discuss again in [5.1.3], by supplying an energy 2Δ, two free electrons are generated by "breaking up" a bound electron pair which can then be scattered. However, the total process is energetically unfavourable for currents below a critical current density, i.e. for a relatively weak shift of the Fermi sphere (in comparison with Δ).

The above considerations require correction if, for more filled bands, the Fermi energy is situated near a band boundary. In the case of two electrons per atom, the valence band is filled and the distribution of occupied states remains unchanged after shifting all occupied states by $\delta \boldsymbol{p}$ (it can easily be seen in the periodic zone scheme). Herewith, a filled band cannot contribute to the total current and to the electrical conductivity (insulator). Only if the considered band is not filled totally, because of an energetic overlap with a neighbouring band inducing a flowing off of electrons into that band, a finite conductivity may be observed.

If both bands do not overlap, but have a small energetic separation of the order of thermal excitation energy $k_B T$, only electrons may be excited by thermal agitation from the filled valence band into the empty conduction band and, therefore, cause a finite conductivity. In this case, we have a thermally activated semiconductor.

The unoccupied states at the top of the valence band are often treated as "holes". The current density $\boldsymbol{j}_\mathrm{h}$ caused by holes equals the total current density of the filled band (equal to zero) minus the current density of the missing electrons $\boldsymbol{j}_\mathrm{e}$

$$\boldsymbol{j}_\mathrm{h} = 0 - \boldsymbol{j}_\mathrm{e} = -en\boldsymbol{v} \tag{11}$$

with n as the corresponding particle density. Consequently, the hole behaves as a positively charged particle. The energy ε_h of a hole is defined appropriately as the missing energy of the electron related to the top of the band ε_0. In its vicinity, the approximation of the effective mass is valid (expansion of $\varepsilon(\boldsymbol{k})$ including quadratic terms in $\boldsymbol{k}$),

$$\varepsilon_\mathrm{e} \approx \varepsilon_0 - \boldsymbol{k}\,\frac{\hbar^2}{2\mathbf{M}^*}\,\boldsymbol{k} \tag{12}$$

with an effective mass tensor $1/\mathbf{M}^*$ with positive eigenvalues. Therefore, we obtain for ε_h

$$\varepsilon_\mathrm{h} = \varepsilon_0 - \varepsilon_\mathrm{e} \approx \boldsymbol{k}\,\frac{\hbar^2}{2\mathbf{M}^*}\,\boldsymbol{k}\,. \tag{13}$$

As a consequence of (13), the hole behaves like a particle with a positive effective mass similar to an electron at the bottom of the band. The behaviour of the electron at the top of the band like a "donkey-particle" against acceleration in an external field, may now be interpreted by the positive electrical charge of the hole. This positive hole charge is also expressed in the sign of circulation in a magnetic field opposite to that of an electron (compare fig. 4.1-3).

4.2 Experimental Determination of the Fermi Surface

4.2.1 Definition of the Fermi Surface

Experimental verification of calculated band structures is important because of the high significances of band structures both for the explanation and evaluation of many macroscopic quantities and properties of a solid, and for a principal understanding of solids. The determination of the Fermi surface makes an important contribution for the solution of this task, because of the accuracy obtainable and because of the simple connection between experimental and calculated quantities. In the last 20 years, the possibilities of experimental determination of the Fermi surface have promoted considerably the development of metal physics. Therefore, we select this field from a series of other problems, and treat it here in more detail.

We remark in this connection that the determination of Fermi surfaces in particular demonstrates a characteristic trend in modern physical research, namely the joined application of different experiment techniques for the investigation of a given system.

The definition of Fermi surfaces is based on the sharp fall of the average occupation number $f(\varepsilon, T)$ from (1.1-7) in Fermi statistics at small temperatures T near the Fermi energy ε_F (compare fig. 4.2-1 for $T = 0$). However, in

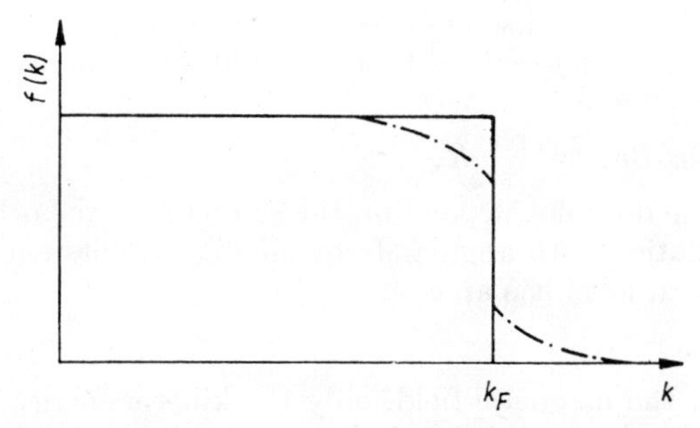

Fig. 4.2-1. Average occupation number $f(k) = f(\varepsilon(\boldsymbol{k}), 0)$ of the homogeneous electron gas as a function of wave number at zero temperature. The broken line indicates the change of $f(k)$ due to the mutual Coulomb interaction. For $K > K_F$ the curve is called correlation or Coulomb tail

real systems, the electrons interact strongly via Coulomb repulsion. The result of this mutual influence is that the particle momentum is no longer a conserved quantity. The probability of finding a momentum state $\hbar \boldsymbol{k}$ occupied can no longer be calculated using (1.1-7). In the ground state of the total system, electrons with momenta higher than the Fermi boundary momentum $\hbar \boldsymbol{k}_F$ may also be observed: the distribution function becomes "smeared out" because of the electron-electron interaction. A more detailed investigation of the many-body problem "interacting electron gas" yields the momentum distribution presented in fig. 4.2-1 by a dot-dashed line (for temperature $T = 0$). The new distribution of the interacting system is characterized again by a step at the same position k_F as for the non-interacting system, which is very important for the definition of the Fermi surface. Therefore, the Fermi surface may now be defined as the isoenergetic surface in $\boldsymbol{k}$-space, showing the step in occupation probability at zero temperature.

4.2.2 Influence of External Magnetic Fields

In the following investigations, we often consider systems under the influence of an external magnetic field. Although the motion of an electron should become more complicated under this additional influence, it has the advantage that the electrons in a magnetic field $\boldsymbol{B}$ move in a plane normal to $\boldsymbol{B}$, which simplifies the problem. In the following, we often limit discussion for simplicity to the example of free electrons. This may be generalized easily for lattice electrons by substituting $\hbar^2 k^2/2m$ by $\varepsilon(\boldsymbol{k})$. A semiclassical treatment of electron motion in a magnetic field has already been given in [4.1] using the effective mass tensor. In this consideration, quantum theory has been used only for calculation of the dispersion relation $\varepsilon(\boldsymbol{k})$ for lattice electrons. Now, we are interested in the quantization of the motion in the magnetic field. According to classical mechanics, electrons in a magnetic field $\boldsymbol{B}$ execute circular and helical motion, respectively, with cyclotron frequency $\omega = |e|\, B/m$. After quantization, these harmonic oscillations lead to a discrete energy spectrum.

The angular frequency is determined by the equilibrium condition of centrifugal force and Lorentz force

$$m\omega^2 R = |e|\, Bv\,, \quad v = \omega R\,, \tag{1}$$

as $\omega = |e|\, B/m$. Here ω does not depend on the radius R of the orbit.

The harmonic oscillation with angular frequency ω exhibits energy quanta $\hbar\omega$, and the n-th excited level has an energy of

$$\varepsilon_n = \hbar\omega n\,. \tag{2}$$

During the motion in the magnetic field, only the kinetic energy $\varepsilon_n = p_n^2/2m$ is present. We obtain for the area of the circle in momentum space corresponding to ε_n, the value $F_n = \pi p_n^2 = 2\pi m \varepsilon_n$ and combining (1) and (2)

$$F_n = |e|\, Bnh\,, \tag{3}$$

(with $h = 2\pi\hbar$), we have the Onsager quantization condition in momentum space. (3) is also valid for more general dispersion relations.

Now we are again able to represent the energy eigenvalues (2) in momentum space as for electrons without a magnetic field. In the plane normal to $\boldsymbol{B}$, we have to consider only the discrete values corresponding to circles with radii p_n. The corresponding allowed cylinders ("Landau cylinders") in momentum space are filled respectively only up to the Fermi energy, as shown in fig. 4.2-2.

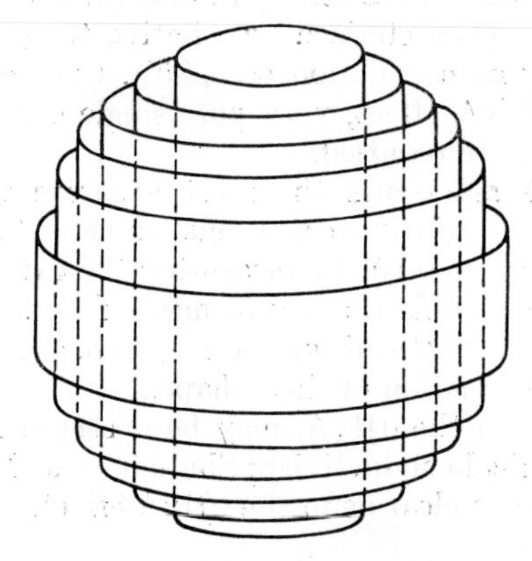

Fig. 4.2-2. Occupied energy states in a magnetic field

4.2.3 De Haas-van Alphen Effect

For magnetic fields currently available in laboratories (1 to 30 T), we obtain in metals approximately 10000 occupied Landau cylinders. According to (3), their cross-section depends on the applied magnetic field. Moving a cylinder across the Fermi surface induces a sharp change in physical quantities only if the cylinder touches the Fermi surface, because at that moment, by a small increase in magnetic field, many electrons are redistributed. This is the reason for many measured quantities to depend periodcally on $1/B$. The separation of two strongly increasing regions follows from (3) as

$$\Delta\left(\frac{1}{B}\right) = \frac{1}{B_n} - \frac{1}{B_{n+1}} = \frac{|e|\,h}{F(\varepsilon_F)} \tag{4}$$

with $F(\varepsilon_F)$ as the extremal cross-section of the Fermi surface normal to the applied magnetic field. These oscillations periodic in $1/B$ are remarkably strong in the magnetic susceptibility χ, and constitute the de Haas-van Alphen (dHvA) effect. The extremal cross-section of the Fermi surface is given by the frequency of the oscillations and yields the main experimental information from the dHvA effect.

So far we have discussed only the case for vanishing temperature $T = 0$. The increase of temperature smears out the step in the Fermi distribution function (1.1-7) and, therefore, reduces the oscillations of χ. Consequently, the amplitude of the dHvA oscillations reduces with increasing temperature. Therefore, for accurate experiments we have to apply low temperatures. (The softening of the Fermi distribution function by $k_B T$ must be small in comparison with the separation of the Landau cylinders $\hbar\omega$, this condition being generally fulfilled for temperatures below 10 K for magnetic fields of the order of 30 T.) Similarly, perturbations of the ideal crystal structure also cause a broadening of the Fermi distribution function characterized by a fictive temperature, the so-called Dingle temperature (for its calculation see [5.5]). Consequently, for accurate measurements of the dHvA effect, very pure single crystals with a minimum of lattice imperfections are required.

At present, the dHvA effect represents the most accurate method for measuring Fermi surfaces (with a relative error smaller than 0.1%!). For closed Fermi surfaces (e.g. alkaline metals), by measuring the extremal cross-sections for all directions of magnetic field the Fermi surface may be constructed unequivocally. For complicated Fermi surfaces, further experimental quantities are necessary for determination of their shape.

Periodic fluctuations in $1/B$ as in the dHvA, may be observed for a series of measured quantities (e.g. electrical conductivity: Shubnikov-de Haas effect) but in no case is the phenomenon so clear as in the dHvA effect.

4.2.4 Cyclotron Resonance

In experiments concerning the cyclotron resonance effect, the angular frequency of electrons in the magnetic field $\omega_c = |e| \, B/m_c^*$ is determined by measuring the frequency-dependent absorption of a high-frequency field. If its frequency coincides with ω_c, the system becomes resonant and the high frequency circuit containing the metal sample to be investigated is strongly damped.

Of course, the cyclotron mass m_c^* does not agree with the free electron mass, but depends on the dispersion relation of the lattice electrons. Since during the absorption process electrons are excited from an occupied to an unoccupied cylinder, we measure the energy difference depending on the change of extremal cross-section F of the Fermi surface with energy:

$$m_c^* = \frac{1}{2\pi} \frac{\partial F}{\partial \varepsilon}. \tag{5}$$

This is tantamount to assertions on the velocity $v = \dfrac{\partial \omega}{\partial k}$ (averaged over the electron orbit). Only those frequencies may be discriminated with which many electrons circulate. This is the case for the extremal cross-sections of the Fermi body, where the Landau cylinders just touch the Fermi surface. At the Fermi surface of fig. 4.1-2b) for the given direction of the magnetic field, this is the case for the "bellies" and "necks" B and N, respectively (see also fig. 4.2-3). Changing the directions of the magnetic field $\boldsymbol{B}$ out of the symmetry direction, the resonances at the necks may vanish at certain angles as indicated in fig. 4.2-3 (Fermi surfaces similar to those in fig. 4.2-3 appear in noble metals, compare fig. 1.2-5).

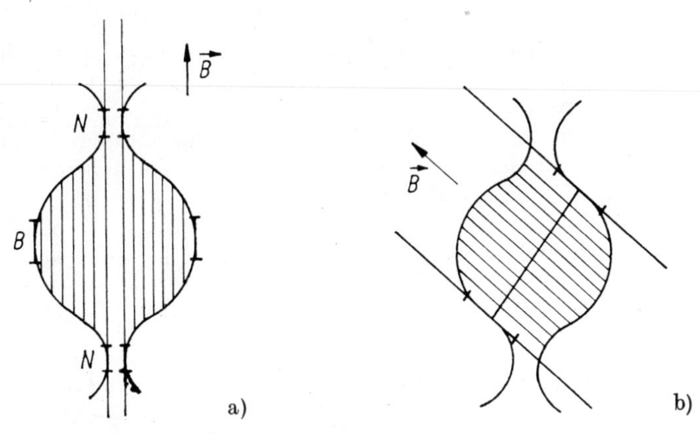

Fig. 4.2-3. a) Extremal cross sections of the Fermi surface of fig. 4.1-2b)
b) Extremal cross section for another direction of $\boldsymbol{B}$

The experimentally determined and theoretically calculated cyclotron masses show in metals an almost constant difference between 10 and 30%. This effect is caused by the electron-phonon interaction (see [5.4]). As a consequence of this strong interaction, the electron energies near the Fermi energy have to be renormalized according to (5.1-6). For the calculation of m_c^*, one must use not the inert mass of the free electron m but the "field mass" induced by the interaction: m_c^* must then be averaged over the corresponding cyclotron orbits. Therefore, so far in cyclotron resonance experiments the cyclotron frequency has not been used for determination of absolute parameters of the Fermi surface, but instead only for relative relations (topology). Especially valuable statements may be made on the disappearing of extremal cross-sections as a function of the direction of $\boldsymbol{B}$ (as shown in figs. 4.2-3a) and b)).

The resonance effect observed for many metals appears only if the electron with circulation frequency ω_c executes many circulations $\omega_c\tau$ in time τ between two scattering events (at defect atoms or lattice vibrations)

$$\omega_c\tau \gg 1. \tag{6}$$

This condition may be fulfilled for large τ. τ increases with the purity (chemical and physical) of the material and with decreasing temperature. Therefore, as in the case of the dHvA effect, the experiments are carried out with single crystals as pure and perfect as possible and at helium temperatures (1.6 to 2 K). Further, (6) may be implemented by large ω. Often, fixed frequencies of the order of 10 GHz are used. Then for molybdenum, for example the required magnetic induction amounts 0.1 to 1 T.

4.2.5 Magneto-Acoustic Effects

A series of magneto-acoustic effects also permits experimental investigation of the structure of the Fermi surface. Generally, these include interaction effects of lattice electrons with phonons (scattering, absorption, emission) under the influence of magnetic fields. Also in this case, the measured quantities (e.g. the absorption coefficient for acoustic waves) depend on $1/B$ periodically which enables a simple geometrical interpretation. For this purpose, we use the fact that the motion of the wave packet in momentum space corresponds to a motion in real space (shifted by a phase $\pi/2$). Of cause, the extension of the wave packet in real and momentum space has to fulfil the uncertainty relation, but the conditions following from it may be neglected in our consideration.

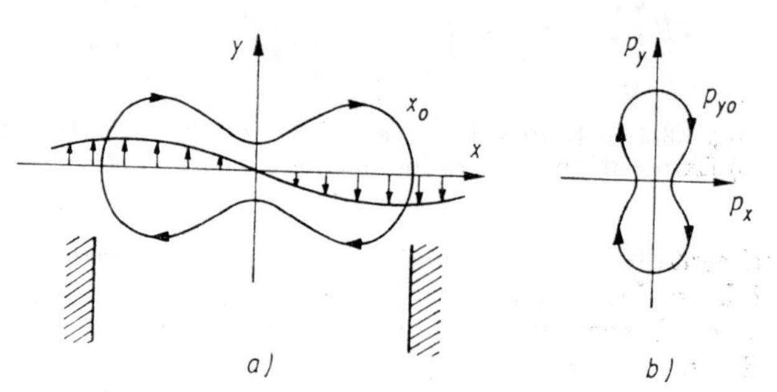

Fig. 4.2-4. a) Orbit of an electron in real space
b) Orbit of the same electron in momentum space

The orbit in real space has the form drawn in fig. 4.2-4a). Then the orbit in momentum space has a form given in fig. 4.2-4b). The deformation of the crystal lattice as a consequence of an acoustic field (in this illustration assumed to be transverse) executes forces on the electron indicated in fig. 4.2-4a) by arrows (x_0 covers many atomic distances). The electron will be accelerated after several circulations if the geometrical resonance condition is fulfilled,

$$\frac{\lambda}{2}(2n+1) = 2x_0 = \frac{2P_{y0}}{|e|\,B}, \tag{7}$$

i.e. if the diameter of the extremal orbit $2x_0$ equals an odd multiple of the half wavelength (in fig. 4.2-4a) the lowest possible value is assumed). This diameter $2x_0$ corresponds to the diameter of the Fermi surface P_{y0} perpendicular both to the direction of $\boldsymbol{B}$ and to the propagation direction of the phonon. Also in this effect, only electrons at the Fermi surface have to be considered since only these can absorb energy. An essential contribution is obtained again only from electrons moving on extremal orbits. Since the oscillation time of the acoustic wave τ_s is large compared with the circulation time τ_c of the electron ($\tau_s \gg \tau_c$), the elongation of the acoustic wave may be assumed as approximately time-independent.

The acoustic field will be weakened by the amount of energy absorbed by the electrons. Consequently, the measurement of the absorption coefficient of the acoustic waves as a function of its propagation direction $\boldsymbol{n}$ (perpendicular to $\boldsymbol{B}$) and of the wavelength gives direct information on the extremal diameters of the Fermi surface (more exactly: on distances of two corresponding parallel planes parallel to $\boldsymbol{B}$ and $\boldsymbol{n}$ and touching the Fermi surface).

A series of other magneto-acoustic effects may be used also for the determination of parameters of the Fermi surface (e.g. also in longitudinal acoustic fields). As the electron-phonon interaction may change strongly over the Fermi surface, the interpretation of the experimental results should be done with caution.

The obtainable accuracy of measurements of the magneto-acoustic effects is not as high as for the dHvA effect. However, the magneto-acoustic effects show the advantage that the geometrical shape of the Fermi surface may be determined more easily than from the dHvA effect. The extremal diameters $2x_0$ of the orbits, discussed above, depend on the strength of the magnetic field. When the sample has just this thickness (in fig. 4.2-4a) the regions outside the sample are hatched), then this may also be observed in the high frequency circuit. By means of this so-called size effect, information on extremal orbital parameters of the Fermi surface may be obtained.

4.2.6. Magneto-Resistance

The increase in electrical resistance in magnetic fields $\boldsymbol{B}$ is denoted magneto-resistance. The behaviour at high fields $\boldsymbol{B}$, pure samples, and low temperatures (or for small electrical resistance or large scattering times τ) is especially interesting because under these conditions, the details of the scattering process become unimportant in comparison with the influence of the shape of the Fermi surface on the conductivity. This condition may be expressed by equ. (6) via the cyclotron frequency ω_c.

Usually, the transverse ($\boldsymbol{j} \perp \boldsymbol{B}$) resistance saturates at high fields (i.e. it assumes constant values). However, if "open" orbits exist (fig. 4.2-5) the resistance increases proportionally with B^2. Open orbits are characterized by the property that they extend in the periodic zone scheme over the total $\boldsymbol{k}$-space. The example in fig. 4.2-5 may be understood as a cut of the Fermi surface from fig. 4.1-2b) with a plane containing the rotational axis. In fig. 4.2-5, the applied field $\boldsymbol{B}$ lies in the z-direction. According to the consideration given above, the electrons may move in real space only in the y-direction (for $B \to \infty$). This means that the electrical resistance in the x-direction becomes infinite.

Therefore, from the direction dependence of the magneto-resistance (for different directions of the applied magnetic field), information on the presence and position of open orbits on the Fermi surface may be obtained.

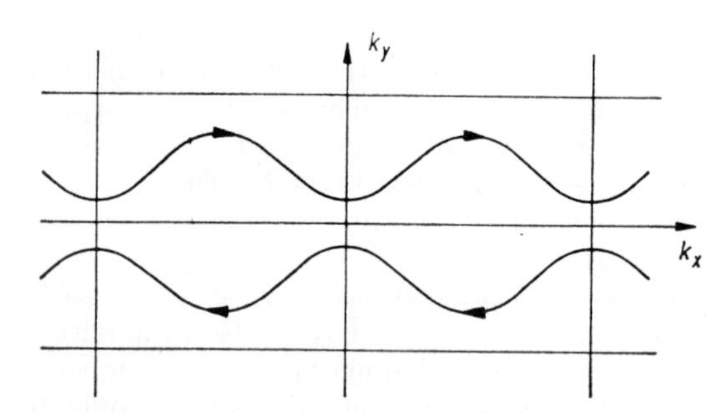

Fig. 4.2-5. Fermi surface with open orbits (see fig. 4.1-2b))

4.2.7 Positron Annihilation

Positron annihilation and the Compton effect represent two experiments for the measurement of cross-sections and chord length of the Fermi body, where low temperatures, very pure crystals, and magnetic fields are not necessary. In the positron annihilation experiment, the radiative decay of a positron-electron pair into two photons (energies $\hbar\omega_{1,2}$, momenta $\hbar\boldsymbol{k}_{1,2}$) is used measuring the angular correlation between the photons. Before the annihilation, the incoming positron is slowed down by inelastic scattering events in approximately 10^{-12} to a nearly vanishing momentum $\boldsymbol{p}_\mathrm{p} = 0$. To describe the effect a relativistic treatment of energy and momentum, a balance is necessary (compare fig. 4.2-6),

$$mc^2 + c\sqrt{(mc)^2 + p_\mathrm{e}^2} = \hbar(\omega_1 + \omega_2), \qquad (8')$$

$$\boldsymbol{p}_\mathrm{e} = \hbar(\boldsymbol{k}_1 + \boldsymbol{k}_2). \qquad (8'')$$

It is obvious that for the implementation of (8), the photon energies should be of the order of the rest energies mc^2 (≈ 0.5 MeV) of the electron. On the other hand, the kinetic energy of the electron with momentum $\boldsymbol{p}_\mathrm{e}$ inside the Fermi sphere is smaller than 10 eV. Therefore, the term p_e^2 in the square root in (8′) may be omitted. The ratio of the momenta p_e and $\hbar k$ is smaller than 5×10^{-3}. The angle ϑ in fig. 4.2-6a) is of the order of this ratio and, therefore, also very small (only several tens of angular minutes).

After eliminating k_2 (via $\omega = ck$), we obtain from (8)

$$\boldsymbol{p}_\mathrm{e} = \hbar(k_1\boldsymbol{e}_1 + k_2\boldsymbol{e}_2) = \hbar k_1(\boldsymbol{e}_1 - \boldsymbol{e}_2) + 2mc\boldsymbol{e}_2. \qquad (9)$$

The allowed values of k_1 for a given angle ϑ according to fig. 4.2-6b) lie on the broken line inside the Fermi body at a distance

$$p_\perp = mc\vartheta \tag{10}$$

from its centre. All electrons on this line contribute to the measured photon intensity at angle ϑ. Therefore, from the intensity of the photons, a conclusion may be drawn on the chord length of the Fermi body belonging to angle ϑ and lying within the plane spanned by the propagation vectors $\boldsymbol{e}_1$ and $\boldsymbol{e}_2$ of the simultaneously emitted photons.

The absolute value of the intensity is determined by the transition probability $W \sim |(\psi_{el}, \exp(i\boldsymbol{kr})\, \psi_{pos})|^2$. Consequently, for its calculation we need information on the wave functions of the electron ψ_{el} and positron ψ_{pos}. As the

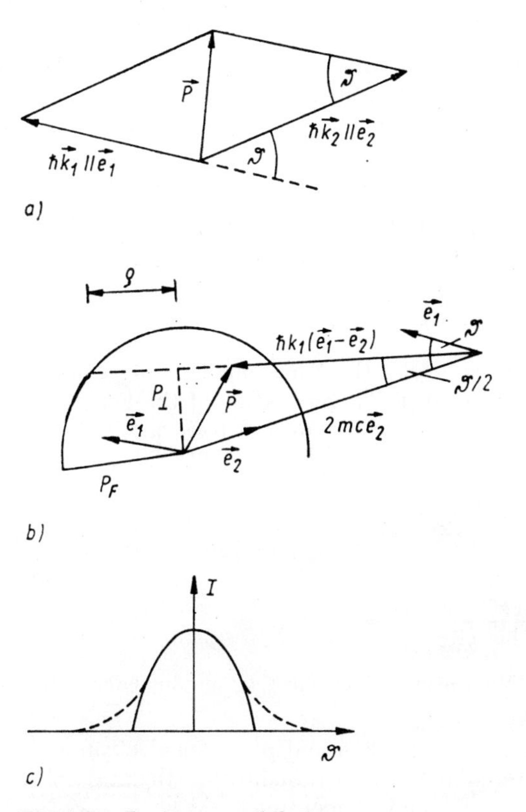

Fig. 4.2-6. Positron annihilation
a) momentum balance
b) possible values of $\boldsymbol{k}_1$ and $\boldsymbol{p}$
c) intensity I as a function of angle of coincidence ϑ

radiated electron just has to compensate the spin of the positron, the use of polarized positrons permits the measurement of Fermi surfaces of a given spin direction (important for magnetically ordered systems).

The annihilation process needs a time of order 10^{-10} s. As sources of positrons, β^+-radiating substances (Na^{22}, Co^{57}, Cu^{46}) are used. In experimental equipment both γ-counters with high angle resolution are arranged in directions $\boldsymbol{e}_1$ and $\boldsymbol{e}_2$ and one measures the number of simultaneously arriving (and therefore resulting from the same radiation process) γ-quanta by a coincidence circuit. If all photons with arbitrary momentum components in a direction perpendicular to $\boldsymbol{e}_1$ an $\boldsymbol{e}_2$ are registered, the cross-section of the Fermi body perpendicular to $(\boldsymbol{e}_1 + \boldsymbol{e}_2)$ and at a distance $p_\perp$ from the origin may be determined immediately. For a spherical Fermi surface of radius p_F, the discussed cross-section F is given by

$$F = \pi\varrho^2 = \pi(p_F^2 - p_\perp^2) = \pi(p_F^2 - m^2c^2\vartheta^2). \tag{11}$$

Consequently, the intensity as a function of ϑ shows a parabola-like form, fig. 4.2-6c), where the experimental result is given by the dotted line.

4.2.8 Compton Effect

The Compton effect is observed by scattering of X-rays with wavelength λ on electrons. Together with scattering through an angle ϑ, they experience an increase in wavelength of

$$\Delta\lambda = \lambda_c(1 - \cos\vartheta), \tag{12}$$

corresponding to an energy loss of the X-ray quantum ($\lambda_c = h/mc$ is referred to as the Compoton wavelength). For investigation of the Fermi body, the spectral intensity of the quanta scattered by the fixed angle ϑ as a function of energy loss is measured (for the exact theory, see [5.2.2]).

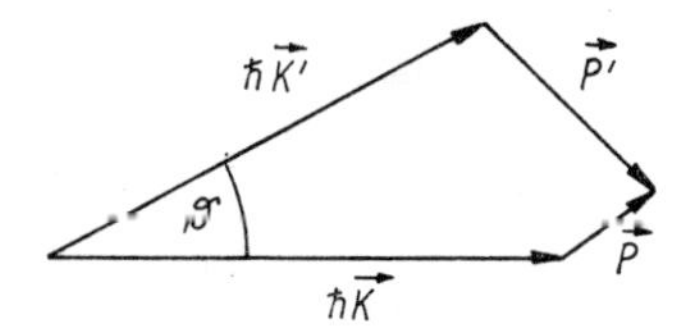

Fig. 4.2-7 Momentum balance for the Compton effect

Figure 4.2-7 shows the momenta $\boldsymbol{p}$ and $\boldsymbol{p}'$ of the electron and $\hbar\boldsymbol{K}$ and $\hbar\boldsymbol{K}'$ of the photon before and after (primed quantities) the scattering event. The conservation laws for energy and momentum yield

$$\sqrt{(mc)^2 + p^2} + \hbar K = \sqrt{(mc)^2 + p'^2} + \hbar K', \tag{13'}$$

$$\boldsymbol{p} + \hbar\boldsymbol{K} = \boldsymbol{p}' + \hbar\boldsymbol{K}'. \tag{13''}$$

In (13') we neglect p^2 compared with $(mc)^2$, eliminate $\boldsymbol{p}'$ by means of (13''), put in the product $\boldsymbol{p}(\boldsymbol{K} - \boldsymbol{K}') = \boldsymbol{p}(\boldsymbol{e} - \boldsymbol{e}')\, K$, and yield from (13) after introducing the frequency $\omega = cK$

$$\boldsymbol{p}(\boldsymbol{e} - \boldsymbol{e}') = mc\,\frac{\hbar\Delta\omega}{\hbar\omega} - \frac{\hbar\omega}{c}\,(1 - \cos\vartheta)\,. \tag{14}$$

Thus, all electrons with the same projection of momentum on the direction $(\boldsymbol{e} - \boldsymbol{e}')$ contribute to the intensity I of quanta scattered through an angle ϑ with energy loss $\hbar\,\Delta\omega$,

$$\boldsymbol{p}(\boldsymbol{e} - \boldsymbol{e}') = p_{\perp}\,|\boldsymbol{e} - \boldsymbol{e}'| = \text{const}\,. \tag{14'}$$

Therefore, the number of these electrons, and with it the intensity, are proportional to the cross-section of the Fermi body with a plane normal to $(\boldsymbol{e} - \boldsymbol{e}')$ and a distance $p_{\perp}$ from the origin. In the case of a spherical Fermi surface, we again obtain for the dependence of I on $\Delta\omega$ parabolae similar to those of fig. 4.2-6c), but now the position of their maximum is given by the energy loss via (12) and via

$$\frac{\Delta\omega}{\omega} = \frac{\hbar\omega}{mc^2}\,(1 - \cos\vartheta)\,, \tag{12'}$$

respectively. For a non-moving electron, scattering is only observed if (12) is fulfilled.

According to (5.2-24), for lattice electrons the contribution of one electron is proportional to the probability

$$W_{\boldsymbol{k}} = \frac{1}{\Omega}\,|(\psi_{\boldsymbol{k}}, \mathrm{e}^{i\boldsymbol{k}\boldsymbol{r}})|^2 \tag{15}$$

of finding the electron described by the wave function $\psi_{\boldsymbol{k}}$ in an eigenstate of momentum. For the derivation of (15), we assumed that the electron after the scattering process may be well described by a plane wave (more exactly: by an OPW) because of its high energy.

Thus, the measurement described gives information on the momentum distribution of lattice electrons. The method exhibits the same advantages as the positron annihilation experiments. Furthermore, for the calculation of the intensity I, a knowledge of the positron wave function is no longer necessary. On the other hand, the Compton effect has the disadvantage that for energy losses high enough ($\hbar\,\Delta\omega > \varepsilon_{\text{core}}$) lower core bands bound with an energy $\varepsilon_{\text{core}}$ also contribute to the measured intensity. Therefore, the energy losses should be restricted to values $\hbar\,\Delta\omega < \varepsilon_{\text{core}}$.

4.3 Photo-Electron and X-Ray Spectroscopy

4.3.1 General Situation

The influence of external radiation on solids gives broad possibilities of their excitation. This includes bombardment with electrons, photons (e.g. ultraviolet light, X-ray quanta, etc.) and others. Either the absorption of this radiation or the emission of particles is measured. Examples for this have already been treated in the last section (Compton effect, positron annihilation). A selection of possible processes occurring within a solid is represented in fig. 4.3-1. For low energies ε_1 there appear very narrow bands resulting from tightly bound atomic states. Higher bands (ε_2) show a small broadening as a consequence of interaction with neighbouring atoms. The conduction bands with energies ε_3 exhibit widths of the order of 10 eV. The states below the Fermi energy ε_F are occupied. Bands above ε_F become broader and broader and, moreover, overlap leading in practice to a continuum of states. Electrons with energies above ε_0 may leave the solid (therefore, in fig. 4.3-1, indicated as free states).

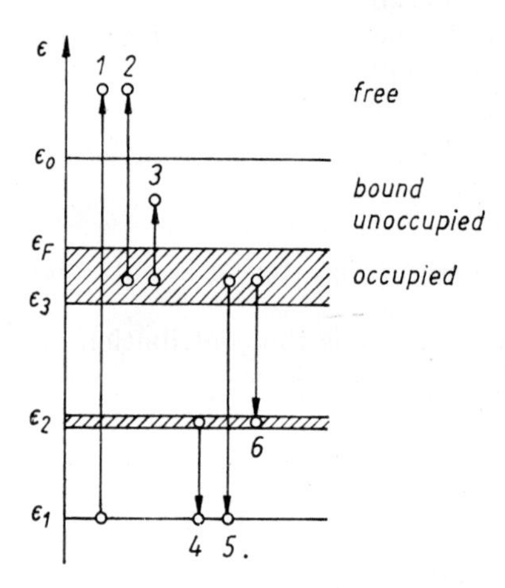

Fig. 4.3-1. Schematic representation of excitation and deexcitation processes in a solid as a consequence of external influence

By energy transfer from the incoming radiation, electrons from occupied states may be excited into empty states. In cases (1) and (2) the electron coming from a low-lying or a conduction band, respectively, may leave the crystal and be analysed with respect to its energy, momentum, and spin. In case (2) the structure of the conduction band can be investigated in this way at a fixed excitation energy delivered by the incident radiation. In case (3) the excited electron is not able to escape from the sample. Therefore, it cannot be measured and one has to investigate the change in energy and momentum of the incident radiation. In cases (4) to (6), electrons in states of higher energy turn to unoccupied low-energy states which have been created by an external influence. In this process, both phonons may be emitted and other electrons may be excited electronically and after that be analyzed (Auger effect). According to the "Golden rule", the intensity of a quantum emitted into a certain solid angle is given by

$$I(\varepsilon, \varepsilon', \boldsymbol{p}, s) \sim \varepsilon \sum_{\mathrm{i(f)}} |\langle \psi_\mathrm{i}| H^I |\psi_f\rangle|^2 \, \delta(\varepsilon + \varepsilon_\mathrm{i} - \varepsilon_\mathrm{f} - \varepsilon'). \tag{1}$$

It is measured as a function of excitation energy ε and depends also on energy ε', momentum $\boldsymbol{p}$, and spin s of the emitted quantum. The right-hand side contains the square of the matrix element of the interaction H^I and describes the probability of the transition from the initial state ψ_i to the final state ψ_f of the system. In these states, all internal (quasi) and external (test) particles participating in the interaction process have to be included. The δ-function ensures the energy balance. If the change of momentum (angular dependence) can be determined by the equipment, then a δ-function for the momentum balance has to be added. Furthermore, one has to average or sum over the possible final (f) and initial (i) states, respectively, according to the particular experimental conditions.

If one of both participating states is known exactly (e.g. the low-lying initial state ψ_i) from I, conclusions may be drawn on the other state (ψ_f). These conclusions concern both the energy ε_f and the number of states available at this energy (i.e. the DOS $D(\varepsilon)$), respectively, and also via the amount of matrix element information on the shape of the wave function. A broadening of the levels as a consequence of their finite life time and of other experimental conditions reduces the accuracy of the information available.

For transitions of type (3) which may be induced, for example, by optical excitation, the energy difference between initial and final state is given, and both vary themselves across the whole band. As a result, so-called mixed DOS are measured containing coupled information on both initial and final states.

4.3.2 X-Ray Emission Spectra

In the following, the emission of X-rays will be discussed in more detail. This corresponds to processes (4) to (6). In (1), the interaction $H^\mathrm{I} = |e| \, \boldsymbol{pA}/m$ with the radiation field is determined by the vector potential $\boldsymbol{A} = \sqrt{\hbar/2\varepsilon_0 \Omega\omega} \boldsymbol{n} \exp(i\boldsymbol{kr})$ with the polarization vector $\boldsymbol{n}$. As the extension of the tightly bound empty state is small in comparison with the wave length of a soft X-ray quantum ($ka \approx 10^{-1} \ldots 10^{-2}$), we can use for H^I the so-called dipole approximation:

$$H^\mathrm{I} = \frac{|e|}{m} \boldsymbol{pA} \sim \omega^{-1/2} \boldsymbol{pn} \, \mathrm{e}^{i\boldsymbol{kr}} \approx \boldsymbol{np}\omega^{-1/2}. \tag{2}$$

Since we consider a spontaneous emission process in (1), the dependence on the energy ε of the radiation incoming on the sample vanishes. We denote the energy of the emitted photon by $\varepsilon' = \hbar\omega$. The matrix element of momentum is replaced by that of the position operator[1]):

$$\langle \psi_\mathrm{i} \, |\boldsymbol{np}| \, \psi_\mathrm{f}\rangle = im\omega \langle \psi_\mathrm{i} \, |\boldsymbol{nr}| \, \psi_\mathrm{f}\rangle. \tag{3}$$

[1]) This relation is a consequence of the commutation relation $[H, \boldsymbol{r}] = \hbar\boldsymbol{p}/im$ and of the fact that ψ_i and ψ_f are eigenstates of H with energies ε_i and ε_f, whereby $\hbar\omega = \varepsilon_\mathrm{i} - \varepsilon_\mathrm{f}$ holds.

If in (3) the matrix elements of the operator $\boldsymbol{r}$ do not depend on the energy, then the fine-structure of the X-ray line according to (1) to (3) gives direct information on the density of occupied states,

$$I(\omega) \sim \omega^2 D(\omega + \varepsilon_f/\hbar) . \tag{4}$$

However, the exact structure will be influenced by the energy dependence of the matrix elements. For simple metals, the matrix elements of the position operator calculated by means of pseudo wave functions agree well with the experiment (better than the matrix elements of the momentum operator). For the spectra of transition metals, the selection rules concerning the angular momenta of initial (l) and final (l') state,

$$l' = l \pm 1 \tag{5}$$

contained in (3) play a decisive role. According to the atomic spectra, the solid spectra are denoted after the shells for the final states of the transition, K, L, M, etc. These tightly bound final states have a well defined angular momentum l'. Because of (5), the contributions of the conduction states belonging to l have to be selected according to the representation of the Bloch function

$$\psi_{\boldsymbol{k}}(\boldsymbol{r}) = \sum_L i^l c_L(\boldsymbol{k})\, R_l(r, \varkappa)\, Y_L(\boldsymbol{r}) \tag{6}$$

inside the Wigner-Seitz cell. Comparison with experiment now yields, instead of (4), the "partial" density of states (m = magnetic quantum number)

$$D_l(\varepsilon) = \sum_{\boldsymbol{k},m} |c_L(\boldsymbol{k})|^2\, \delta(\varepsilon - \varepsilon_{\boldsymbol{k}}) . \tag{7}$$

The energy dependence of the residual integrals via the radial wave functions may be neglected in most cases for transition metals.

Using polarized X-rays and single crystals makes available information on the orientation of the wave function via the direction dependence of the matrix elements (3).

For heavy transition elements ($Z > 40$), the spin-orbit interaction of the states has a strong influence on the shape of the spectra (Nemoshkalenko et al. 82). The split of the d-states as a consequence of this interaction has the same order as the width of the d-band. For the electronic structure of actinides, see Weinberger (84), (90).

As the low-lying states of a certain type i of atom are well coordinated to their energy, the energy dependent matrix element tests the widely extended states of the conduction band precisely at the site of atom $\boldsymbol{R}_{\mathrm{i}}$, since the wave function of the final state $\psi_f(\boldsymbol{r} - \boldsymbol{R}_{\mathrm{i}})$ is well localized at that site (partial "local" DOS). Of course, very low-lying states are less informative because of large level broadening as a consequence of their reduced life time.

Interpreting the experimental spectra, many-particle effects (e.g. plasmon excitations) have also to be taken into account. An important problem arises from the question of the extent to which the Coulomb interaction of the hole state influences the one-electron function of the Bloch state. Recently, it was shown that the final state of the process without low-lying hole states plays the decisive role: consequently, the matrix element may be calculated using ordinary Bloch wave functions (von Barth, Grossmann 79, Mahan 80). However, a strict proof on the basis of many-particle theory of strongly bound hole states is still an open question.

4.3.3. Photo-Electron Spectroscopy

With the availability of tunable high-power radiation sources such as synchrotron radiation sources, photo-electron spectroscopy (PES), in the ultraviolet (UPS), and in the X-ray spectral region (XPS) have given increasing importance to the experimental investigation of the electronic structure of solids. Thereby, the incoming light quantum with energy $\varepsilon = \hbar\omega$ excites, corresponding to processes (1) and (2) in fig. 4.3-1, an electron which will be transported in a second step to the surface of the sample and will escape from the solid in a third step, undergoing reflection and scattering effects at the surface. As a consequence of the strongly inelastic processes during the second step with UPS, only electrons from the uppermost layers may be analysed, whereas for XPS with electron energies of about 1 keV with an electronic mean free path of the order of 5 to 10 nm bulk effects are certainly seen. The dipole approximation (2) is also applicable for the excitation process.

For UPS, the change in momentum of the electrons is small because of the small momentum of the incident photon. Therefore, we obtain the "perpendicular" transitions represented in fig. 4.3-2. Generally speaking, the inter-

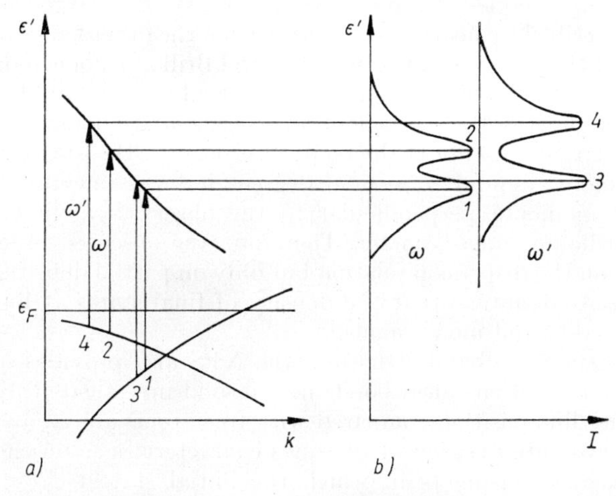

Fig. 4.3-2. Explanation of the kind of information obtainable by emission of photoelectrons

pretation of the spectra is the more complicated, the more unknown states are involved in the transition probabilities. Advantages arise for PES because the external conditions (such as the energy of the incoming radiation) may be varied widely as shown schematically in fig. 4.3-2. In fig. a) a band structure with two crossing occupied bands is seen.

Incoming photons with frequency ω induce transitions (1) and (2). This causes the shape of the intensity as a function of energy ε' of the emitted electron as represented in fig. 4.3-2b). The transitions (1) and (2) are broadened to the two neighbouring maxima (1 and 2). For a somewhat increased incoming energy $\hbar\omega' > \hbar\omega$, transitions (3) and (4) result. As a consequence of the opposed tendencies of both crossing bands, the distance between the maxima (3) and (4) of the intensity distribution function (for the incoming energy $\hbar\omega'$) changes strongly in comparison with the change $\hbar(\omega - \omega')$. Therefore, by a change of ω a variety of valuable information may be gained. The interpretation is facilitated if the position of a "reference band" is known. This is the case if dispersionless surface states are present. In the angle-resolved PES (ARPES), the direction of the $\boldsymbol{k}$-vector is given by the direction of the observed emitted electron. Therefore, it is thus possible to measure directly the dispersion relation of occupied and unoccupied bands (Eastman, Himpsel 81). From the height of the intensity maxima, information on the matrix elements may be gained. Using polarized radiation enables determination of symmetry properties of the bands. The analysis of the spins of the emitted electrons allows assertions on spin-polarized bands in magnetically ordered systems (ferromagnetic exchange splitting in Co: Eastman et al. 80). These measurements give indications that the exchange splitting evaluated with the potential V_{xc} (1.3-13) of von Barth and Hedin (72) becomes too large.

In the XPS, the energies of the excited electrons are so large that the final state may be described by nearly free electrons. On the other hand, the momentum change is of the order of the diameter of the Brillouin zone and, therefore, cannot be neglected. The diffraction of the emitted electron at the surface of the sample complicates the treatment of energy and momentum balance. Because of the large momenta of the emitted electrons, the angular resolution of the spectrometer limited to several degrees induces uncertainties in the components of momenta perpendicular to the observation direction of the order of the Brillouin zone diameter. Therefore, the observed intensity gives no information on the dispersion relation but only on partial densities of states. Moreover, it is of advantage that the density of final states at high energies changes only weakly and monotonously.

Besides statements on band structure, the XPS also provides information on charge relations and chemical bonds in the solid investigated by using the shifts in absolute line position compared with free atoms and known standard materials. The absolute position of the lines characterizes as do characteristic X-rays, the chemical elements in which the initial state is localized. As a consequence, the PES is often also denoted ESCA (Electronic Spectroscopy for Chemical Analysis) and used for chemical analysis (Siegbahn 67).

The intensity of the XPS photoelectrons from core levels show a strong angular dependence (ARXPS). It is caused by the scattering of the photoelectron at the atoms in the neighbourhood of the emitter. The evaluation of the spectra gives information on the lattice spacing, the surface reconstruction, the position of adsorbate atom etc. — To explain ARXPS it is necessary to go beyond the plane wave approximation for the final state. It is a sum of spherical waves with an origin at the emitter and at the neighbouring atoms. The single scattering process can be described by the plane wave approximation, but more refined considerations include curved wave and multiple scattering corrections (Rennert and Hung 88).

5 Further Electronic Properties

5.1 Density of States, Heat Capacity, Transition Temperature of Superconductivity, Magnetic Susceptibility, and Transport Quantities

The DOS at the Fermi energy, D_F, essentially determines the electronic heat capacity. The electron-phonon interaction modifies this relation, causes superconductivity, and determines its transition temperature. D_F and the electron-electron interaction determine the susceptibility of the Pauli spin paramagnetism. Transport quantities such as electrical resistivity can be calculated by solving the Boltzmann equation; the latter contains transition probabilities, which are determined by the band structure and the phonon spectrum, respectively.

5.1.1 Density of States and Heat Capacity

The DOS $D(\varepsilon)$ follows from the band structure $\varepsilon_{\boldsymbol{k},\nu}$ as the number of electron states per energy interval and unit cell according to (cf. 1.2-24 as well)

$$D(\varepsilon) = \frac{2}{N} \sum_{\substack{\boldsymbol{k}\in \mathrm{BZ} \\ \nu}} \delta(\varepsilon - \varepsilon_{\boldsymbol{k},\nu}) = \frac{2\Omega_0}{(2\pi)^3} \sum_{\nu} \int_{\mathrm{BZ}} \mathrm{d}^3k \delta(\varepsilon - \varepsilon_{\boldsymbol{k},\nu}) . \tag{1}$$

The factor 2 takes into account the two spin directions. Each of the N unit cells has a volume Ω_0. The Fermi energy is determined by the requirement that the integral over $D(\varepsilon)$ with ε_F as the upper bound just equals the total number of electrons per unit cell.

The partial information about the band structure contained in $D(\varepsilon)$ determines the equilibrium thermodynamics of the electron system via

$$U_{\mathrm{el}}(T) = N \int \mathrm{d}\varepsilon D(\varepsilon)\, f(\varepsilon, T)\, \varepsilon\,, \tag{2}$$

which contributes to the internal energy $U(T) = U_{\mathrm{el}}(T) + U_{\mathrm{ph}}(T)$. $U_{\mathrm{el}}(T)$ contains with $f(\varepsilon, T)$ the Fermi-Dirac distribution function (1.1-7). The contribution of the phonon system, $U_{\mathrm{ph}}(T)$, being analogous to (2), contains the Bose-Einstein distribution function (5.3-13) according to (5.3-14). As for $U(T)$, the heat capacity at constant volume Ω

$$C_\Omega(T) = \big(\partial U(T)/\partial T\big)_\Omega \tag{3}$$

also consists of two parts, $C_\Omega(T) = C_{\mathrm{el}}(T) + C_{\mathrm{ph}}(T)$. At low temperatures T ($\ll \theta_D$, the Debye temperature), it holds that $C_{\mathrm{el}}(T) \sim T$ and $C_{\mathrm{ph}}(T) \sim T^3$, and thus (for non-magnetic metals)

$$C_\Omega(T) = \gamma T + AT^3 + \cdots. \tag{4}$$

The term proportional to T is caused by the smearing out of the Fermi edge. It follows from the evaluation of $\partial U_{\text{el}}/\partial T$ with $U_{\text{el}}(T)$ from (2) that at low T, $\partial f(\varepsilon, T)/\partial T$ essentially differs from zero only in a small interval $\sim k_B T$ around the Fermi energy ε_F (see fig. 1.1-1). There $D(\varepsilon)$ can be considered constant and can be replaced by a constant factor $D_F = D(\varepsilon_F)$, which simplifies the evaluation of the integral: $\gamma \sim D_F$. The evaluation of the remaining integral yields the proportionality factor

$$\gamma = \frac{\pi^2}{3} k_B^2 D_F. \tag{5}$$

Furthermore, the term proportional to T^3 in (4) describes the contribution of acoustic phonons (see [5.3.2]). Via the Debye energy $k_B\theta_D = \hbar\omega_D$, A contains the atomic density (of the metal) and the mean sound velocity (of the acoustic phonons). Plotting $C_\Omega(T)/T$ versus T^2 (see fig. 5.1-1), a straight line with intercept γ and slope A is obtained. Thus γ and A, and consequently microscopic quantities D_F and θ_D or ω_D, can be determined from low-temperature measurements of $C_\Omega(T)$; the comparison of values determined theoretically and experimentally showing a discrepancy, which can be explained by the fact that the influence of the lattice oscillations on the electron states has been neglected in this consideration.

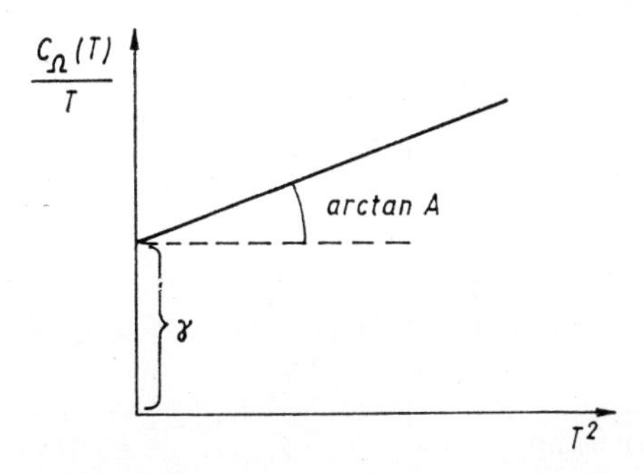

Fig. 5.1-1. Heat capacity of non-magnetic metals at low temperatures

5.1.2 Renormalization of Electron Energies

In the preceding sections, only the interaction of an electron with the static crystal potential and with the other electrons was taken into account. The corrections to the adiabatic approximation discussed in [1.3] lead to the electron-phonon interaction. According to [1.3], this interaction particularly modifies the behaviour of the electrons in the vicinity of the Fermi level, $|\varepsilon_{\boldsymbol{k},\nu} - \varepsilon_F| \sim \hbar\omega_D$. It can even lead to an effective attraction, which is the basis of the phonon mechanism of superconductivity. In an obvious physical picture, an electron (via Coulomb forces) polarizes the lattice in its surroundings. This distortion can in its turn act on other electrons. In a dynamic picture, the strength of the interaction is determined by the phonon energies of order $\hbar\omega_D$, which can be exchanged during this process.

The dynamic properties of an electron are also changed thereby, since the electron is compelled to drag its lattice polarization (or, in another representation, the cloud of virtual phonons it produces) with itself. This can best be described by a $\boldsymbol{k}$-dependent renormalization of the dispersion relation

$$\varepsilon_{\boldsymbol{k},\nu} \to \varepsilon_{\boldsymbol{k},\nu}/(1 + \lambda_{\boldsymbol{k},\nu}). \tag{6}$$

Analogously, the DOS at the Fermi level proves to be

$$D_F \to D_F(1 + \lambda) \quad \text{with} \quad \lambda = \sum_{\boldsymbol{k},\nu} \lambda_{\boldsymbol{k},\nu}\delta(\varepsilon_{\boldsymbol{k},\nu} - \varepsilon_F)/\sum_{\boldsymbol{k},\nu} \delta(\varepsilon_{\boldsymbol{k},\nu} - \varepsilon_F) \tag{6'}$$

as the averaged electron-phonon coupling parameter. Its calculation is explained in [5.4]. It can be determined by means of tunnel spectroscopy or from the temperature T_c of the transition into the superconducting state. The orders of magnitude of D_F and λ for some metals are found in table 5.1-1.

The factor $(1 + \lambda)$ should also be taken into account in the electronic part of the heat capacity

$$\gamma = \frac{\pi^2}{3} k_B^2(1 + \lambda) D_F. \tag{5'}$$

Table 5.1-1. DOS D_F at the Fermi energy, electron-phonon coupling parameter λ, transition temperature of superconductivity T_c, and Debye energy $\hbar\omega_D$ for selected metals. The data for λ_{emp} are optimum values taken from tunneling and T_c-measurements after McMillan (68), Grimvall (81), Butler (77). λ_{theor} for simple metals see Grimvall (81); for transition metals: Papaconstantopoulos et al. (77) for $3d$; Butler (77) for $4d$; John et al. (82) for $5d$. For simple metals, the corresponding r_s-values (see [1.1.2]) are given in brackets.

metal	D_F/ryd^{-1}	λ_{theor}	λ_{emp}	T_c/K	$\hbar\omega_D$/meV
Be (1.87)	0.80	0.26	0.24	0.03	89
Zn (2.29)	4.28	0.42	0.43	0.85	20
Pb (2.29)	7.10	1.47	1.55	7.19	9
α-Hg (2.65)	4.08	0.93	1.60	4.16	9
V	25.40	1.05	0.6	5.30	34
Nb	20.60	1.39	0.97	9.26	22
Ta	21.62	1.16	0.69	4.48	19

Particularly for the transition metals with their tendency to form long-range magnetic order, one has to remember that in the λ figuring in (5′) a contribution caused by spin fluctuations is also contained, besides the electron-phonon part. This seems to apply particularly to the beginning of the transition period with its tendency to anti-ferromagnetic order (Sc, Y, Lu) and to the end, with the tendency to ferromagnetic order (Pd, Pt). The influence of the spin fluctuations on the electronic heat capacity can be suppressed by applying an external magnetic field.

The renormalization of the electron energies also manifests itself in other measurable quantities as cyclotron masses and amplitudes of dHvA oscillations. Quantities connected with the geometry of the Fermi surface only, such as the periods of dHvA oscillations, remain unchanged.

5.1.3 Transition Temperature of Superconductivity

The effective attraction of electrons near the Fermi energy brought about by electron-phonon interactions — provided that it exceeds the Coulomb repulsion of electrons there — leads to the formation of so-called Cooper pairs from electrons with opposite momenta and antiparallel spins. According to the theory of Bardeen, Cooper, and Schrieffer (BCS), all electrons in the vicinity of the Fermi level in a range with a width of the Debye energy $\hbar\omega_D$ condense into Cooper pairs. As distinct from $\varepsilon_F \approx 5 \cdots 10$ eV, the Debye energies amount to as little as some 10 meV: for specific examples for T_c and $\hbar\omega_D$, see table 5.1-1. To generate two free electrons, such a Cooper pair with a binding energy of 2Δ has to be broken up. Thus a forbidden zone of this width appears in the one-particle excitation spectrum. If these excitation energies are not available, a state of the system with total momentum different from zero and, thus, an electric current cannot be damped by the scattering of electrons, i.e., a superconducting state is present. With increasing temperature, the energy gap $\Delta(T)$ diminishes until it vanishes at the critical temperature T_c. Then the system takes on the normally conducting state. According to that BCS theory, for the energy gap $\Delta(0)$ at $T = 0$, it holds

$$2\Delta = 3.52 k_B T_c \tag{7}$$

and

$$k_B T_c = 0.86 \hbar\omega_D \, e^{-1/D_F V} . \tag{8}$$

Thus T_c increases with the magnitude $\hbar\omega_D$ of the exchanged energy. It also increases with the number of electrons participating in the pair formation in the vicinity of the Fermi level ($\sim D_F$), as well as the strength of the effectively attracting interaction V, which has been assumed in (8) to be given within a model and to be small.

The improved theory of Eliashberg (60) starts from the system of coupled equations for the Green functions of electrons and phonons. Its approximate numerical evaluation according to McMillan (68) yields the expression

$$k_B T_c = \frac{\hbar\langle\omega\rangle}{1.20} \exp\left[-\frac{1.04(1+\lambda)}{\lambda - \mu^*(1+0.62\lambda)}\right] \tag{9}$$

which is also valid for stronger interactions ($\lambda \approx 1$). Here $\langle\omega\rangle$ is a mean phonon frequency. The Coulomb pseudopotential μ^* globally takes into account the repulsive Coulomb interaction of the conduction electrons (for transition metals it is assumed to be $\mu^* \approx 0.13$, for simple metals somewhat smaller values are found). (8) is contained in (9) as a special case for small values of λ and $(\lambda - \mu^*)$.

Of the pure metals, Pb, α-Hg, Nb, and V (see table 5.1-1) exhibit maximum values of λ, whilst for most other metals λ is about 0.5. For the transition metals and their alloys, T_c shows a strong dependence on the number Z of valence electrons per atom (Matthias rule). Maxima of T_c are found for $Z = 4.5$ and 6.75. This tendency could be confirmed by calculating the dependence of λ along the $3d$- and $4d$-periods (Papaconstantopoulos et al. 77, Butler 77) as well as the $5d$-period (John et al. 82) of the transition metals. Interpolation in the spirit of a rigid-band model explains the values for the alloys. Altogether these data show that high values of D_F do not necessarily correspond to high values of λ as shown by the examples of Pd and Pt, which show only small λ and thus no superconductivity in spite of a high D_F. At most, it holds vice versa that small values of D_F are coupled with small values of λ. To this end, we consider the elements Ta ($Z = 5$) and W ($Z = 6$). Both $5d$-metals have the same lattice structure (*fcc*). The band structures and DOS are very much alike besides the fact that ε_F lies near a maximum of $D(\varepsilon)$ for Ta, while lying in a minimum for W ($D_F^W = 5$ per atom and *ryd*, $D_F^{Ta} = 22$ per atom and *ryd*). The corresponding values of λ and T_c are $\lambda^W = 0.3$, $\lambda^{Ta} = 0.7$, $T_c^W = 0.012$ K, $T_c^{Ta} = 4.48$ K. Analogous statements hold for the corresponding $4d$-metals Nb ($Z = 5$, $\lambda = 0.97$, $T_c = 9.26$ K) and Mo ($Z = 6$, $\lambda = 0.44$, $T_c = 0.92$ K).

An important test for the theory is the pressure dependence of T_c. While for simple metals the influence of the pressure dependence of the phonon spectrum dominates and leads to $\partial T_c/\partial p < 0$, the positive values of $\partial T_c/\partial p$ observed at the beginning of the transition metal periods can be explained with the help of the theory treated in [5.4]. The physical explanation is based on the relatively large ionic radii of these elements.

Ferromagnetism as a cooperative phenomenon with different band structures for different spin directions competes with the superconductivity and its balance of electron spins in the Cooper pairs (see [3.2]). The possibility of their coexistence has been in dispute for a long time (Vonsovskij et al. 77). Therefore, recently there has been interest in ternary rare-earth compounds[1])

[1]) Examples are (RE)M_4B_4 with for example M = Rh, Ru, Ir, Os and RE = Er, Y; (RE)Mo_6S_8 with for example RE = Ho; and (RE)Rh_xSn_y, $Er_3Rh_4Sn_{13}$; $(RE)_2Fe_3Si_5$, $(RE)_5Co_4Si_{10}$, (RE)Cu_2Si_2. In these compounds the $4f$-electrons of the RE ions carry the magnetic moments (localized at the corresponding lattice sites), whereas the $3d$-, $4d$-, and $5d$-electrons of the transition metal ions are mainly responsible for the super-

as well as in the binary compound Y_9Co_7[1]) as "magnetic superconductors".[2])[3]) For UPt_3, one observes the coexistence of spin fluctuations and superconductivity. For Pd, also a sensitive interaction between both these influences is to be observed. On the one hand, as a consequence of a very narrow $4d$-band the magnetic susceptibility is very high, which leads to strong spin fluctuations. On the other hand, the calculated λ is not as small as to exclude superconductivity at all. While Pd tends to ferromagnetic order, there is a tendency to antiferromagnetic order with Y. The competition between ferromagnetism and superconductivity is also found with the lanthanides.

5.1.4 Superconductors with High Critical Temperatures

After a long period of stagnation (Nb_3Ge — the compound with the highest T_c of 23.2 K — was already known in 1973!), Bednorz and Mueller (86) discovered a new class of superconductors with considerably higher critical temperatures T_c: The importance of this break-through was recognized by the awarding of a Nobel prize in physics only one year after the first publication! Currently, three groups of high-temperature superconductors are known (fig. 5.1-2):

(1) $(La, M)_2CuO_4$ with M = Ca, Sr, Ba (fig. b))
 (K_2NiO_4 structure type), highest T_c = 35 K[4]) for $La_{1.85}Sr_{0.15}CuO_{4-y}$,
(2) $MBa_2Cu_3O_{7-y}$ with M = Sc, Y, La, Nd, Sm, Gd, Tb, Dy, Ho, Er, Tu, Yb and binary combinations with Lu; $y = 0 \cdots 0.5$ (denoted as 123 structure, fig. c)), $T_c \approx 95$ K (Wu et al. 87),
(3) $Bi_2Sr_2Ca_nCu_{n+1}O_{n+6+y}$; T_c = 110 K[4]) ($n = 2$), 85 K ($n = 1$), <20 K ($n = 0$), ($Tl_2Ca_2Ba_nCu_{n+1}O_{n+6+y}$; T_c = 127 K[4]) ($n = 2$), 105 K ($n = 1$), 80 K ($n = 0$), in these alloys perovskite and BiO layers alternate (see fig. d)).

They are oxidic systems with a perovskite-like structure which show metallic conductivity and a non-zero partial DOS of oxygen at the Fermi energy.

A common structure element of all three types is a CuO_2 layer (fig. 5.1-3) with a partially filled band which is formed by antibonding bonds between Cu $3d$-states of symmetry $(x^2 - y^2)$ and O $2p$-states in the x- or y-direction. The role of the other layers is still unclear. Probably, they are important (i) as a polarizable medium which determines the lattice constant and separates the

conductivity. There are peculiarities of the superconductivity in $CeCu_2Si_2$ and UBe_{13} (heavy fermion systems). For further details see Eckern et al. (84).

[1]) Here magnetic order arises from itinerant electrons.

[2]) With the organic superconductors $(TMTSF)_2X$, X = ClO_4, PF_6, AsF_6, antiferromagnetic order appears above the transition temperature.

[3]) For the theory of magnetic superconductors see Tachiki (83).

[4]) The data on T_c differ somewhat because the transition curves of the electrical resistivity as a function of temperature have a width of a few K depending on the homogeneity of the samples.

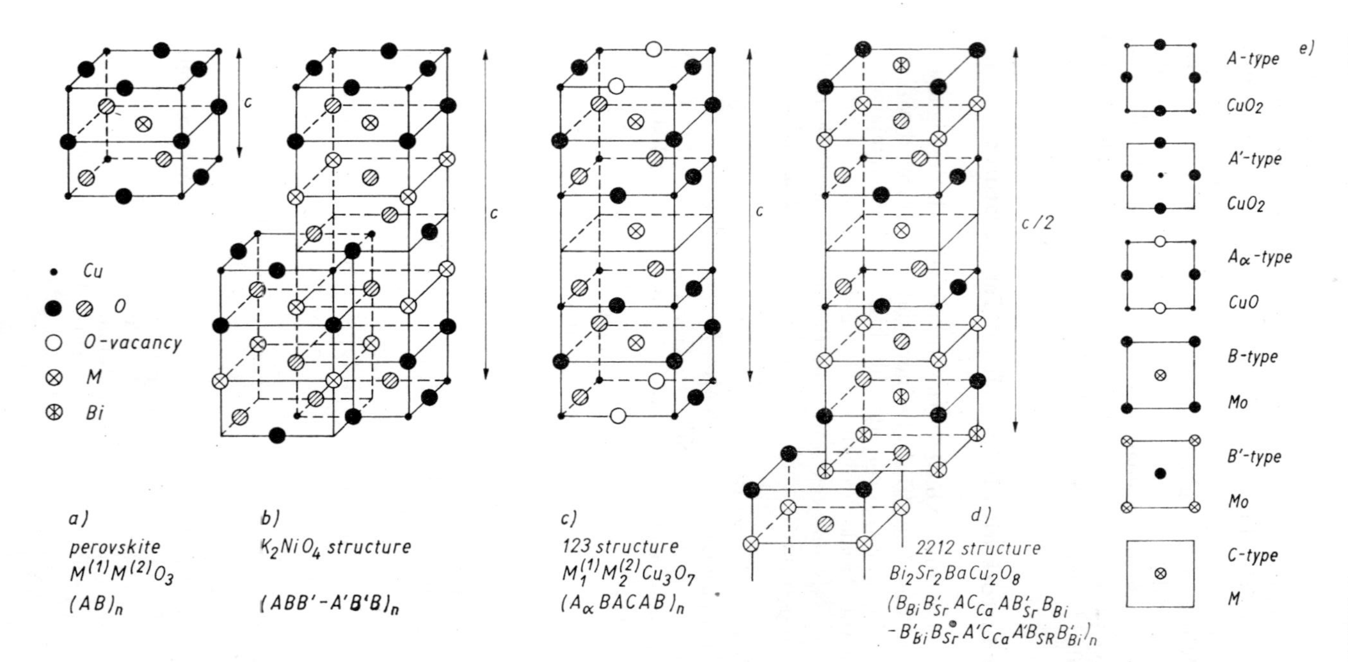

Fig. 5.1-2. Lattice structure of perovskite-like high-temperature superconductors. They result by stacking the different layers A, A′, A_α, B, B′, or C of fig. e). For simplicity the distances between the layers are assumed to be equal (c lattice constant in stacking direction)
a) perovskite structure b) K_2NiO_4 structure type, the M sites are occupied by La, Sr, Ba, or Ca atoms, after 3 layers the stack is shifted by (1/2, 1/2, 0) in the direction of the layer diagonal c) 123 structure, the layer A_α differs from the layer A by the missing oxygen atoms; therefore, it consists of Cu—O chains in y-direction d) 2212 structure, the CuO_2 layers are separated by Ca layers alternating with a group of SrO—BiO—BiO—SrO layers e) types of elementary layer A and A′ or B and B′, respectively, differ by a shift of half the diagonal (1/2, 1/2, 0) only. Type A_α results from A by the marked oxygen vacancies. C layers consist only of metal atoms without oxygen

CuO_2 layers, and (ii) for doping. T_c and other physical properties depend very strongly on oxygen content (denoted by the deviation y from the stoichiometry in the structure formulae). In particular, defects (vacancies) in the CuO chains in layers of type A_α determine the filling of the Cu $3d$—O $2p$ band. The layer structure is responsible for strong anisotropies in electrical conductivity and in the parameters of superconductivity.

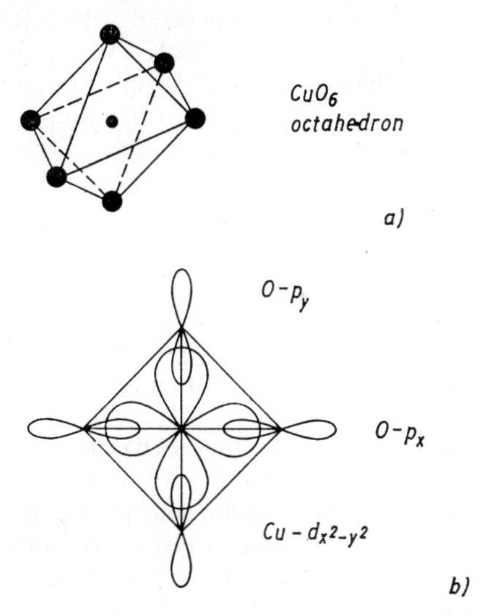

Fig. 5.1-3. a) CuO_6 octahedron as an element of short range order in perovskite-like structures. In the 123 and 2212 structure, one or two oxygen atoms are missing but the Cu atom is always surrounded by 4 atoms like in fig. b).
b) Bonding between $3d_{x^2-y^2}$ and $2p_x$, $2p_y$ states of the neighbouring Cu and O atoms, respectively

The mechanism of superconductivity is also still unclear. It has been shown experimentally that electron pairs are responsible for superconductivity. In comparison to known metallic systems, the non-vanishing DOS for oxygen sites at the Fermi energy is an important new feature. It implies that the oscillations of the (light!) oxygen atoms contribute to the electron-phonon coupling parameter λ in (5.4-9). Because of the relation $\langle\omega\rangle \sim M^{-\beta}$, a dependence on the oxygen mass (isotope effect) with exponent $\beta = 1/2$ is expected, according to (5.4-9). The observed isotope effect is much weaker ($\beta = 0.03\cdots0.05$ in 123 structures, Leary et al. 87). Therefore, the conclusion may be drawn that other, non-phononic mechanisms also contribute to λ. Other new features of the high-T_c superconductors to be taken into account in discussing the electron-phonon mechanism are the anisotropy and anharmonicty of the lattice vibrations. Quantitative estimations of these effects show that T_c of the K_2NiO_4 structure may be understood under these assumptions, but for the explanation of the experimental results in the 123 structures, a strong additional electronic correlation in the narrow Cu $3d$—O $2p$ band (with a band width $\approx 3\cdots5$ eV, smaller than in the known superconducting systems) is generally assumed. At present, a strong controversy on the actual form of these contributions exists.

Apart from a basic point of view, the new high-T_c superconductors are of great practical importance because they can be operated with liquid nitrogen as the cooling medium (cooling temperature $T_N = 77$ K), which reduces costs considerably. Usually it is assumed that for technical applications the critical temperature T_c should exceed the cooling temperature by factor 1.5 at least (i.e. $T_c \gtrsim 115$ K).

To this aim, the following problems must be solved:

- processing of flexible wires or tapes from brittle oxidic materials,
- increasing the density of the critical current of the material produced by ceramic technology (presently 10^3 A/cm^2) above values of 10^5 A/cm^2.

By means of molecular beam epitaxy, very thin crystalline layers with critical current densities of $1.5 \cdot 10^7$ A/cm^2 have been obtained. The observed critical magnetic fields H_c exceed considerably those of other known superconductors.

Several publications have reported superconductors with critical temperatures at or above room temperature. However, the results are not accepted by the physical community because either the superconducting properties are stable only for a short time, or else the experiments are not at all reproducible.

Enormous further development of this field is to be expected.

5.1.5 Magnetic Susceptibility

Another quantity determined by D_F is the magnetic susceptibility. The contributions to this quantity are: diamagnetism of closed-shell ion cores, Pauli spin paramagnetism of valence electrons, and Landau orbital diamagnetism of valence electrons.

Let us consider first spin paramagnetism. Its calculation by Pauli in 1927 proved to be the first application of quantum statistics to solid state physics. It results from the magnetic spin momenta of the electrons being adjusted themselves in an external magnetic field $\boldsymbol{B}$. The two possibilities for the spin are connected with energy shifts

$$\Delta\varepsilon = \pm\mu_s B \quad \text{for} \quad \begin{cases} \text{spin antiparallel to } \boldsymbol{B} \\ \text{spin parallel to } \boldsymbol{B} \end{cases}$$

related to the zero-field case (note that $\mu_s < 0$ because of $e < 0$: spin and magnetic momentum are opposite to each other). Therefore, in the case of free electrons, the band with $\varepsilon_{\boldsymbol{k}}^0 = (\hbar \boldsymbol{k})^2/2m$ and $D^0(\varepsilon) = D_F^0 \sqrt{\varepsilon/\varepsilon_F^0}$, $D_F^0 = 3/2\varepsilon_F^0$ is replaced by two bands shifted about $\pm\mu_s B$ with one half of the original DOS each and different spin directions (see fig. 5.1-4). Both bands are filled up to the common Fermi level at $T = 0$. The surplus electrons with momenta parallel to the field therefore equals the doubly hatched area

$$N^0_{\pm} = \int_{\pm\mu_s B}^{\varepsilon^0_F} d\varepsilon \, \frac{1}{2} \, D^0(\varepsilon \mp \mu_s B) \curvearrowright$$

$$N^0_+ - N^0_- = \int_{\varepsilon^0_F + \mu_s B}^{\varepsilon^0_F - \mu_s B} d\varepsilon \, \frac{1}{2} \, D^0(\varepsilon) \approx |\mu_s| \, B D^0(\varepsilon^0_F) \, . \tag{10}$$

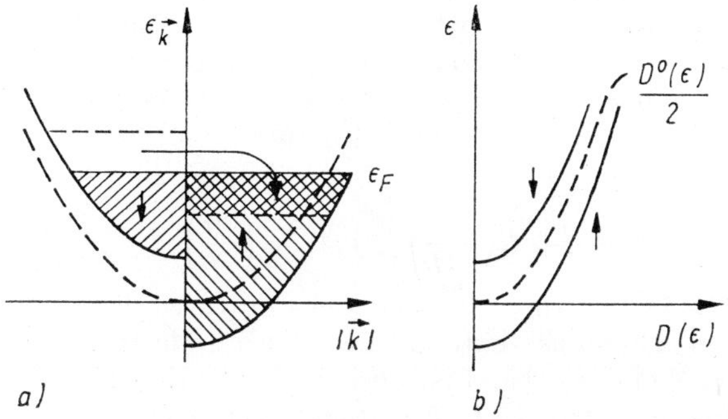

Fig. 5.1-4. Free electrons in a magnetic field, a) energy states, b) DOS

For the magnetization $M^0 = (N^0_+ - N^0_-) \, |\mu_s|/\Omega$, it therefore follows that $M^0 = \chi^0_{\text{para}} \, B$ with

$$\chi^0_{\text{para}} = \mu_s^2 \, \frac{D^0(\varepsilon^0_F)}{\Omega} \tag{11}$$

as the spin susceptibility. Thus it is (as is the electronic heat capacity) proportional to the DOS at the Fermi level. An analogous relation holds for Bloch electrons $\chi_{\text{para}} = \mu_s^2 D(\varepsilon_F)/\Omega$. As a consequence of Fermi-Dirac statistics, χ_{para} is nearly independent of temperature. Incidentally, χ_{para} can also be determined directly from nuclear resonance measurements (Knight shift).

Finally the orbital diamagnetism explained by Landau in 1930 yields $\chi^0_{\text{dia}} = -(1/3) \, \chi^0_{\text{para}}$ (the negative sign results from the fact that the switching-on of the magnetic field is connected with an energy increase for the electrons), free electrons thus behaving paramagnetically altogether and contributing $(2/3) \, \chi^0_{\text{para}}$ to the magnetic susceptibility. For Bloch electrons the theory of orbital diamagnetism is more complicated. An additional effect is the above-mentioned influence of the ion cores. This is why measurements of susceptibility are less appropriate for the determination of D_F. Furthermore, one has to take into account the renormalization of the susceptibility by the electron-electron interaction (see Moruzzi et al. 78, Ashcroft and Mermin 76)[1]).

5.1.6. Transport Quantities

Many theoretical investigations of transport coefficients, such as electrical and thermal conductivity, and also thermopower, magnetoresistivity (transverse, longitudinal), Hall coefficient, etc. start from the quasiclassical concept of the Boltzmann equation, describing the system of valence electrons by a one-partial distribution function $f(\boldsymbol{r}, \boldsymbol{p}, t)$ or $f_{\boldsymbol{k}}(\boldsymbol{r}, t)$, which follows from the unperturbed Fermi-Dirac distribution (1.1-7) under the influence of small perturbations, which may depend on time and, weakly, on position. The simultaneous information on position $\boldsymbol{r}$ and momentum $\boldsymbol{p}$ is not in contradiction with quantum mechanics since we can assume that the wave packets — as discussed in [4.1] — are much smaller than the regions in which the external perturbations and thus $f_{\boldsymbol{k}}(\boldsymbol{r}, t)$ change to any extent. As mean particle density in phase space, $f_{\boldsymbol{k}}(\boldsymbol{r}, t)$ is normalized to the total number of valence electrons.

The Boltzmann equation reads

$$\left(\frac{\partial}{\partial t} + \boldsymbol{v}_{\boldsymbol{k}} \frac{\partial}{\partial \boldsymbol{r}} + \boldsymbol{F}_{\boldsymbol{k}} \frac{\partial}{\partial \hbar \boldsymbol{k}}\right) f_{\boldsymbol{k}}(\boldsymbol{r}, t) = \left(\frac{\partial f_{\boldsymbol{k}}}{\partial t}\right)_{\text{coll}} \tag{12}$$

with $\boldsymbol{v}_{\boldsymbol{k}} = \partial \varepsilon_{\boldsymbol{k}}/\partial \hbar \boldsymbol{k}$ as the electron velocity according to (4.1-3) and $\boldsymbol{F}_{\boldsymbol{k}} = e(\boldsymbol{E} + \boldsymbol{v}_{\boldsymbol{k}} \times \boldsymbol{B})$ as the external force on an electron. For the sake of simplicity, only one energy band is considered and, therefore, the band index is dropped, hence $k = (\boldsymbol{k}, \nu) \to \boldsymbol{k}$. Equation (12) proves to be a balance equation in phase space, its left hand side (multiplied by $\mathrm{d}^3r\, \mathrm{d}^3p\, \mathrm{d}t$) means the total change of the number of electrons in the phase space volume $\mathrm{d}^3r\, \mathrm{d}^3p$ during $\mathrm{d}t$, if this volume moves with the phase space points according to $\boldsymbol{v}_{\boldsymbol{k}}$ and $\boldsymbol{F}_{\boldsymbol{k}}$. For a volume $\mathrm{d}^3r\, \mathrm{d}^3p$ fixed in phase space, $\mathrm{d}^3r\, \mathrm{d}^3p\, \mathrm{d}t\, (\boldsymbol{v}_{\boldsymbol{k}} \partial/\partial \boldsymbol{r} + \boldsymbol{F}_{\boldsymbol{k}} \partial/\partial \boldsymbol{p}) f_{\boldsymbol{k}}$ is the number of particles which leave this volume through its surface owing to their unperturbed motion in phase space. The change of the particle number due to interaction ('collisions'), e.g. with defects, lattice vibrations, or other electrons, which is important for the evolution towards the equilibrium state, is described by the r.h.s., the so-called 'collision term', which contains the transition probabilities (i. e. the probabilities per time) $P_{\boldsymbol{k},\boldsymbol{k}'}$, for scattering from $\boldsymbol{k}$ to $\boldsymbol{k}'$ (or vice versa because of microscopic reversibility), and, in its turn, describes a balance:

$$\left(\frac{\partial f_{\boldsymbol{k}}}{\partial t}\right)_{\text{coll}} = \sum_{\boldsymbol{k}'} P_{\boldsymbol{k},\boldsymbol{k}'}(f_{\boldsymbol{k}'} - f_{\boldsymbol{k}}), \tag{13}$$

[1]) On the contrary there is no renormalization of χ^0_{para} due to the electron-phonon interaction. The one-particle states are renormalized according to (6) within a range of $\hbar\omega_D$ on both sides of the Fermi energy. This shell 'rides' on top of the Fermi sea and can therefore not be 'filled up' with electrons that turn their spins under anapplied magnetic field (see Grimvall 81, p. 140).

i.e. increase in number of particles in the state $\boldsymbol{k}$ through scattering from all possible states $\boldsymbol{k}'$ to $\boldsymbol{k}$ (first term) minus decrease in particle number referring to $\boldsymbol{k}$ through scattering from $\boldsymbol{k}$ to $\boldsymbol{k}'$ (second term). For given $\boldsymbol{v}_{\boldsymbol{k}}$, $\boldsymbol{F}_{\boldsymbol{k}}$, and $P_{\boldsymbol{k},\boldsymbol{k}'}$, (12) and (13) yield an integro-differential equation to determine $f_{\boldsymbol{k}}(\boldsymbol{r}, t)$. Thus a transport theory based on (12) requires: calculation of the band structure and the quantities following from it; calculation of the transition probabilities; and solution of the Boltzmann equation.

If, for the calculation of the electrical conductivity, we restrict ourselves to an electrical d.c. field as an external perturbation, then it is $\boldsymbol{F}_{\boldsymbol{k}} = e\boldsymbol{E}$, and the first term in (12) equals zero. For weak perturbations, there will be only small deviations $g_{\boldsymbol{k}}(\boldsymbol{r}, t)$ from the Fermi-Dirac equilibrium distribution $f_{\boldsymbol{k}}(\boldsymbol{r}) = f(\varepsilon, T)$: $f_{\boldsymbol{k}} = f_{\boldsymbol{k}}^0 + g_{\boldsymbol{k}}$. The distribution $f_{\boldsymbol{k}}^0$ may depend weakly on the position via $\zeta(\boldsymbol{r})$ and $T(\boldsymbol{r})$. The smallness of $g_{\boldsymbol{k}} = f_{\boldsymbol{k}} - f_{\boldsymbol{k}}^0$ allows us to linearize eqns. (12) and (13) with respect to $\boldsymbol{E}$. Furthermore, if one considers homogeneous materials and fields only, the second term in (12) does not occur either. Finally, with the restriction on elastic scattering processes (scattering on defects only) with $\partial f_{\boldsymbol{k}}^0/\partial \boldsymbol{k} = (\partial f_{\boldsymbol{k}}^0/\partial \varepsilon_{\boldsymbol{k}})\, \partial \varepsilon_{\boldsymbol{k}}/\partial \boldsymbol{k}$ and $f_{\boldsymbol{k}'} - f_{\boldsymbol{k}} = g_{\boldsymbol{k}'} - g_{\boldsymbol{k}}$ (because of $f_{\boldsymbol{k}'}^0 = f_{\boldsymbol{k}}^0$) (12) and (13) take the simple form

$$\frac{\partial f_{\boldsymbol{k}}^0}{\partial \varepsilon_{\boldsymbol{k}}} \boldsymbol{v}_{\boldsymbol{k}} e\boldsymbol{E} = \sum_{\boldsymbol{k}'} P_{\boldsymbol{k},\boldsymbol{k}'}(g_{\boldsymbol{k}'} - g_{\boldsymbol{k}}). \tag{14}$$

The derivative of the Fermi distribution function proves to be a smeared out δ-function with width $k_{\mathrm{B}}T$, which is different from zero only in the vicinity of ε_{F}: $-\partial f_{\boldsymbol{k}}/\partial \varepsilon_{\boldsymbol{k}} = \delta_{k_{\mathrm{B}}T}(\varepsilon_{\boldsymbol{k}} - \varepsilon_{\mathrm{F}})$. Therefore, scattering processes take place only near the Fermi surface. (14) suggests the ansatz

$$g_{\boldsymbol{k}} = \delta_{k_{\mathrm{B}}T}(\varepsilon_{\boldsymbol{k}} - \varepsilon_{\mathrm{F}})\, \boldsymbol{\Lambda}_{\boldsymbol{k}} e\boldsymbol{E} \tag{15}$$

with vectors $\boldsymbol{\Lambda}_{\boldsymbol{k}}$ on the Fermi surface still to be determined, which have the dimension of length. (15) transforms (14) into

$$-\boldsymbol{v}_{\boldsymbol{k}} = \sum_{\boldsymbol{k}'} P_{\boldsymbol{k},\boldsymbol{k}'}(\boldsymbol{\Lambda}_{\boldsymbol{k}'} - \boldsymbol{\Lambda}_{\boldsymbol{k}}) \quad \text{or} \quad \boldsymbol{\Lambda}_{\boldsymbol{k}} = \tau_{\boldsymbol{k}}(\boldsymbol{v}_{\boldsymbol{k}} + \sum_{\boldsymbol{k}'} P_{\boldsymbol{k},\boldsymbol{k}'} \boldsymbol{\Lambda}_{\boldsymbol{k}'}). \tag{16}$$

$1/\tau_{\boldsymbol{k}} = \sum_{\boldsymbol{k}'} P_{\boldsymbol{k},\boldsymbol{k}'}$ being the probability for a particle in the state $\boldsymbol{k}$ to change over into any other state $\boldsymbol{k}'$ at the Fermi surface through (elastic) scattering. For given $\boldsymbol{v}_{\boldsymbol{k}}$ and $P_{\boldsymbol{k},\boldsymbol{k}'}$, (16) is an inhomogeneous linear integral equation for $\boldsymbol{\Lambda}_{\boldsymbol{k}}$, whose solution gives $g_{\boldsymbol{k}}$ according to (15). (It can be solved, for example, with a variational ansatz; also see [5.5], where use is made of the fact that the integral kernel is degenerate; a modern method is based on the use of Fermi surface harmonics, see Mertig (82).)

Let us assume that (16) has been solved and, thus, $g_{\boldsymbol{k}}$ is known. Then, for the electrical current density (being the product of charge density and velocity), it follows that

$$\boldsymbol{j} = \frac{e}{\Omega} 2 \sum_{\boldsymbol{k}} f_{\boldsymbol{k}} \boldsymbol{v}_{\boldsymbol{k}} = \frac{e}{\Omega} 2 \sum_{\boldsymbol{k}} g_{\boldsymbol{k}} \boldsymbol{v}_{\boldsymbol{k}} = e2 \int \frac{\mathrm{d}^3 k}{(2\pi)^3}\, g_{\boldsymbol{k}} \boldsymbol{v}_{\boldsymbol{k}}, \tag{17}$$

where in the second step it has been taken into account that $f_{\boldsymbol{k}} = f^0_{\boldsymbol{k}} + g_{\boldsymbol{k}}$ and that $f^0_{\boldsymbol{k}}$ describes a state with zero current. Inserting (15) and subsequently comparing the euqation with Ohm's law, for $T \to 0$ one obtains the conductivity tensor (for the explanation of $\circ$ see p. 80, footnote 2)

$$\boldsymbol{\sigma} = \frac{1}{4\pi^3} \frac{e^2}{\hbar} \iint\limits_{\varepsilon_{\boldsymbol{k}}=\varepsilon_{\mathrm{F}}} \frac{\mathrm{d}S_{\boldsymbol{k}}}{v_{\boldsymbol{k}}} \boldsymbol{v}_{\boldsymbol{k}} \circ \boldsymbol{\Lambda}_{\boldsymbol{k}}. \tag{18}$$

For cubic symmetry, the conductivity tensor is proportional to the unit tensor, $\boldsymbol{\sigma} = \sigma \mathbf{1}$ with

$$\sigma = \frac{1}{3} \frac{1}{4\pi^3} \frac{e^2}{\hbar} \iint\limits_{\varepsilon_{\boldsymbol{k}}=\varepsilon_{\mathrm{F}}} \frac{\mathrm{d}S_{\boldsymbol{k}}}{v_{\boldsymbol{k}}} \boldsymbol{v}_{\boldsymbol{k}} \boldsymbol{\Lambda}_{\boldsymbol{k}}. \tag{19}$$

Once $\boldsymbol{v}_{\boldsymbol{k}}$ (from band structure calculation) and $\boldsymbol{\Lambda}_{\boldsymbol{k}}$ (from the solution of (16)) are known, σ, or $\varrho = 1/\sigma$, follows from the Fermi surface integration (19).

An important simplification consists in the so-called relaxation time approximation. If one describes the scattering process according to $\boldsymbol{\Lambda}_{\boldsymbol{k}} = \boldsymbol{v}_{\boldsymbol{k}}\tau$ by a scalar and $\boldsymbol{k}$-independent relaxation time τ (this is possible, strictly speaking, only for a spherical Fermi surface — then it also holds that $\boldsymbol{v}_{\boldsymbol{k}} = v_{\mathrm{F}}\boldsymbol{e}_{\boldsymbol{k}}$ and $\boldsymbol{\Lambda}_{\boldsymbol{k}} = \Lambda \boldsymbol{e}_{\boldsymbol{k}}$ with $\Lambda = v_{\mathrm{F}}\tau$), then from (19) in the effective mass approximation with $\boldsymbol{v}_{\boldsymbol{k}} = \hbar\boldsymbol{k}/m^*$ and $4\pi k_{\mathrm{F}}^3/3 = 4\pi^3 n$, it follows that

$$\sigma = \frac{e^2 n}{m^*} \tau \quad \text{or} \quad \varrho = \frac{m^*}{e^2 n} \frac{1}{\tau} \tag{20}$$

(for free electrons m^* is equal to m). For the collision term (13), the relaxation time approximation yields the following expression:

$$\left(\frac{\partial f_{\boldsymbol{k}}}{\partial t}\right)_{\mathrm{coll}} = \sum_{\boldsymbol{k}'} P_{\boldsymbol{k},\boldsymbol{k}'}(g_{\boldsymbol{k}'} - g_{\boldsymbol{k}}) = -\frac{g_{\boldsymbol{k}}}{\tau}. \tag{21}$$

The latter in connection with (12) shows that the collision term in relaxation time approximation provides a physically reasonable exponential decay of an originally existing perturbation of the equilibrium state. On the microscopic level, τ is the mean time of flight. This is also why $\Lambda = v_{\mathrm{F}}\tau$ is the mean free path, usually large in comparison with the mean atomic distance a. Now, we ask, what is τ to be determined from? Inserting $\boldsymbol{\Lambda}_{\boldsymbol{k}} = \boldsymbol{v}_{\boldsymbol{k}}\tau$ into (16) yields

$$\frac{1}{\tau} = \sum_{\boldsymbol{k}'} P_{\boldsymbol{k},\boldsymbol{k}'}(1 - \cos\vartheta), \quad \vartheta = \sphericalangle(\boldsymbol{k}, \boldsymbol{k}'). \tag{22}$$

Here in the sum $\sum_{\boldsymbol{k}'} P_{\boldsymbol{k},\boldsymbol{k}'}\boldsymbol{v}_{\boldsymbol{k}'}$ only the component $(\cos\vartheta)\,\boldsymbol{v}_{\boldsymbol{k}}$ of $\boldsymbol{v}_{\boldsymbol{k}'}$, which is parallel to $\boldsymbol{v}_{\boldsymbol{k}}$, is taken into account. For the isotropic case, which we consider, $P_{\boldsymbol{k},\boldsymbol{k}'}$ depends on the angle ϑ between $\boldsymbol{k}$ and $\boldsymbol{k}'$ only. Thus, in this case τ is indeed independent of $\boldsymbol{k}$. However, (21) is also used for general (e.g. non-elastic) scattering processes and anisotropic Fermi surfaces in the spirit of a semi-empirical approximation.

The transport factor $(1 - \cos\vartheta)$ is proportional to the momentum change of the electron in the direction of $\boldsymbol{v_k}$. Since it is plausible, this factor provides a greater contribution to the considered scattering process, the larger its scattering angle ϑ and thus its momentum change. With (20) and (22), the calculation of electrical resistivity is reduced to the calculation of relaxation time from the transfer probabilities, the latter being the decisive task of transport theory.

Finally, within the relaxation time approximation (21), the generalization to a.c. fields $\boldsymbol{E} \sim \exp(-i\omega t)$ leads to the tensor of a.c. conductivity

$$\boldsymbol{\sigma}(\omega) = \frac{1}{4\pi^3}\frac{e^2}{\hbar}\iint\limits_{\varepsilon_{\boldsymbol{k}}=\varepsilon_{\mathrm{F}}} \frac{\mathrm{d}S_{\boldsymbol{k}}}{v_{\boldsymbol{k}}}\,\frac{\boldsymbol{v_k}\circ\boldsymbol{v_k}}{\dfrac{1}{\tau} - i\omega} \tag{23}$$

instead of (18). (For the stationary solution, this is $g_{\boldsymbol{k}}(t) \sim \exp(-i\omega t)$, which leads to $1/\tau \to 1/\tau - i\omega$ because of $\partial/\partial t \to -i\omega$ according to (12) and (21).) For spherical Fermi surfaces, it follows that $\boldsymbol{\sigma}(\omega) = \sigma(\omega)\,\mathbf{1}$ with

$$\sigma(\omega) = \frac{\sigma(0)}{1 - i\omega\tau}, \quad \sigma(0) = \sigma \tag{24}$$

instead of (20). According to the kind of scattering processes, τ is usually energy and temperature dependent (with optical frequencies, it also depends on the frequency). The properties of τ result from the transfer probabilities of the scattering processes according to (22).

For very low temperatures scattering at chemical impurities (substitutional and interstitial impurity atoms) and deviation from the ideal crystalline structure (voids, interstitial atoms, dislocations, grain boundaries) prevail. These influences lead to the temperature-independent residual resistivity; its first-principle calculation for point defects is described in [5.5].

At higher temperatures, scattering preferably takes place at thermal lattice defects, the phonons. At high temperatures ($T > \theta_{\mathrm{D}}$), the scattering probability is proportional to the quadratic displacement of atoms from the average positions $\overline{u^2}$, and thus to their mean potential energy. Thus according to the classical equipartition theorem, $\varrho_{\mathrm{ph}} \sim T$. The more precise analysis of the electron-phonon scattering process for all temperatures gives the Bloch-Grueneisen formula

$$\varrho_{\mathrm{ph}} \sim \frac{1}{\tau_{\mathrm{ph}}} \sim \left(\frac{T}{\theta_{\mathrm{D}}}\right)^5 \int\limits_0^{\theta_{\mathrm{D}}/T} \frac{\mathrm{d}z\,z^4}{\mathrm{e}^z - 1}. \tag{25}$$

Here the phonon system is simplifyingly described by the Debye model (see [5.3.2]). There is first of all success in calculating the phonon part of the electrical resistivity from first principles (band structure, phonon spectrum, electron-phonon coupling), for example, for Nb and Pd with the method of Fermi surface harmonics (see Pinski et al. 81, also Butler 81).

According to the addition of transition probabilities in $1/\tau$ (see (22)) within the relaxation time approximation the (temperature-independent) defect part and the (temperature-dependent) phonon part of the a.c. resistivity are also to be added (Matthiessen rule). Deviations from the Matthiessen rule result, for examples, from the differences in the anisotropy (i.e. $\boldsymbol{k}$-dependence) of the several scattering processes. In these cases, the relaxation time approximation fails. As a consequence of the excitation of internal magnetic degrees of freedom, magnetical impurities may lead to a peculiarity in the temperature dependence. Instead of increasing continuously with T, ϱ shows a characteristic minimum at low temperatures (Kondo effect).

A complete description of the kinetics of the electron-phonon system requires starting from a coupled system of Boltzmann equations for electrons and phonons. Such an approach allows the reaction of the electron subsystem to the phonon subsystem (phonon drag) to be described. This reaction leads to deviations from Bloch low-temperature behaviour $\varrho_{\mathrm{ph}} \sim T^5$, as follows from (25).

Analogously to the electrical resistivity, other transport quantities can similarly be calculated. Besides electric and magnetic fields, external perturbations may also have the form of inhomogeneous temperature distributions, which lead to heat current densities $\boldsymbol{j}_{\mathrm{h}} \sim \partial T(\boldsymbol{r})/\partial \boldsymbol{r}$.

Finally, it should be mentioned that there are approaches alternative to the Boltzmann equation for the calculation of transport quantities, for instance the current-current correlation functions (Kubo formalism) and force-force correlation functions (for this see, for example, Röpke and Ziesche 77, Möbius et al. 79), which for metals have mostly been evaluated in the lowest order of perturbation so far, equivalent to solving the Boltzmann equation (e.g. by a variational ansatz). A transport theory beyond the validity of the Boltzmann equation, however, must depend on the alternatives mentioned.

This is the case for strong scattering, as occurs in transition metals, with strong electron-phonon interaction, in concentrated alloys, and in amorphous metals. For concentrated alloys, experimental results (e.g. negative temperature coefficients, which actually are in contradiction with metallic behaviour) could be explained satisfactorily. Besides this, for strong disorder — mean free path is comparable to mean atomic distance a ($\Lambda \gtrsim a$) — there is the concept of the minimal metallic conductivity (Mott 76). According to this, if the disorder exceeds a certain measure there exist only localized states, which contribute to the conductivity through phonon-induced hopping processes. Thus, the band conduction has to be replaced by a concept of hopping conductivity to be described with a master equation in this case. (It is not clear yet, whether the saturation behaviour of the resistivity to be observed in many cases can be described satisfactorily in the framework of the Boltzmann equation.) A collective charge transport by charge density waves (Fröhlich mechanism) is found in the case of the quasi one-dimensional conductor $NbSe_3$ (see [6.6]).

With a sufficiently strong external field non-linear transport behaviour occurs; small sizes or strong magnetic fields may make the quasiclassical description fail; many-particle effects of the electron correlation may require particular care (possibly description by one-particle functions does not suffice either). Such a non-linear, fully quantum-mechanic theory of charge transport including correlation effects is required for the description of the processes in sub-micron semiconductor structures.

Recently, peculiarities of the charge transport in thin wires at low temperatures have been discussed. These peculiarities are connected with the circumstance that in strictly one-dimensional systems, all states are localized if there are perturbations present, however small they are. Yet it could be proved that at low temperatures localization occurs in thin wires as well, if the resistivity exceeds a critical value. Then the resistivity is not proportional to the wire length L, but increases exponentially with L (Thouless 77, Giordano 80, Weller and Ziesche 84, 87). Another subject of recent investigations are quantum wires.

5.2 Dielectric Function, Inelastic Electron and Photon Scattering, and Optical Reflection

The inelastic scattering of electrons and photons and the optical reflection result from the dielectric function (see Ziesche and Paasch 83; and Taut 83).

5.2.1 Dielectric Function

An important key quantity for the understanding of several electronic properties of metals is the dielectric matrix of the metal electrons, $\varepsilon(\boldsymbol{q}, \boldsymbol{q}'; \omega)$. A non-local material quantity, within a microscopic description it conveys the connection between an external field $\boldsymbol{E}_{\text{ext}}$ and the microscopic total field $\boldsymbol{E}_{\text{tot}}$ it generates in the solid. We consider an external field $\boldsymbol{E}_{\text{ext}}(\boldsymbol{r}, t)$ depending on position and time. This field can be written as a superposition of temporally and spatially periodic parts with amplitudes $\boldsymbol{E}_{\text{ext}}(\boldsymbol{q}, \omega)$:

$$\boldsymbol{E}_{\text{ext}}(\boldsymbol{r}, t) = \sum_{\boldsymbol{q}} \int \frac{\mathrm{d}\omega}{2\pi\Omega} \, \mathrm{e}^{i(\boldsymbol{q}\boldsymbol{r} - \omega t)} \boldsymbol{E}_{\text{ext}}(\boldsymbol{q}, \omega) \,. \tag{1}$$

$\boldsymbol{E}_{\text{ext}}$ polarizes the metal, the induced polarization charge density en_{ind} (n_{ind} being the induced electron density) gives rise to the field $\boldsymbol{E}_{\text{ind}}$ and thus causes a screening of the external perturbation field $\boldsymbol{E}_{\text{ext}}$. Compared with $\boldsymbol{E}_{\text{ext}}$, $\boldsymbol{E}_{\text{ind}}$ has a spatial periodicity due to the crystal structure (Bragg reflection). $\boldsymbol{E}_{\text{tot}} = \boldsymbol{E}_{\text{ext}} + \boldsymbol{E}_{\text{ind}}$ is the total field. Now, a microscopic material quantity $\varepsilon(\boldsymbol{q}, \boldsymbol{q}'; \omega)$ is introduced according to $E_{\text{tot}} = \varepsilon^{-1} E_{\text{ext}}$ (written in symbols) or

$$E_{\text{tot}}(\boldsymbol{q}, \omega) = \sum_{\boldsymbol{q}'} \varepsilon^{-1}(\boldsymbol{q}, \boldsymbol{q}'; \omega) \, E_{\text{ext}}(\boldsymbol{q}', \omega) \tag{2}$$

(written in full), which just describes the above-mentioned screening of the external field by n_{ind} or $\boldsymbol{E}_{\text{ind}}$, respectively. Along with this, the fields are assumed to alter only slowly (i.e. quasistatically) and thus can be derived from electrostatic potentials $u_{\text{ext}}, u_{\text{ind}}, u_{\text{tot}}$: $\boldsymbol{E}(\boldsymbol{r}, t) = -\partial u(\boldsymbol{r}, t)/\partial \boldsymbol{r}$. As regards the Fourier components, this means $\boldsymbol{E}(\boldsymbol{q}, \omega) = \boldsymbol{e}_{\boldsymbol{q}} E(\boldsymbol{q}, \omega)$, i.e. the fields are longitudinal. Therefore, from the dielectric tensor $\boldsymbol{\varepsilon}(\boldsymbol{q}, \boldsymbol{q}'; \omega)$ introduced by $\boldsymbol{D} = \varepsilon_0 \boldsymbol{\varepsilon} \boldsymbol{E}_{\text{tot}}$ or

$$\boldsymbol{D}(\boldsymbol{q}, \omega) = \varepsilon_0 \sum_{\boldsymbol{q}'} \boldsymbol{\varepsilon}(\boldsymbol{q}, \boldsymbol{q}'; \omega) \, \boldsymbol{E}_{\text{tot}}(\boldsymbol{q}', \omega) \tag{3}$$

only the purely longitudinal component

$$\varepsilon(\boldsymbol{q}, \boldsymbol{q}'; \omega) = \boldsymbol{e} \, \boldsymbol{\varepsilon}(\boldsymbol{q}, \boldsymbol{q}'; \omega) \, \boldsymbol{e}', \quad \boldsymbol{q} = q\boldsymbol{e}, \quad \boldsymbol{q}' = q'\boldsymbol{e}', \tag{4}$$

is required. In (3), which is known from electrodynamics, $\boldsymbol{E}_{\text{tot}}$ is determined by all charges (external and induced), whereas the field $\boldsymbol{D}$ has only the external charges as sources. In (2), $\varepsilon^{-1}(\boldsymbol{q}, \boldsymbol{q}'; \omega)$ denotes the matrix inverse of $\varepsilon(\boldsymbol{q}, \boldsymbol{q}'; \omega)$ with respect to $\boldsymbol{q}, \boldsymbol{q}'$: $\sum_{\boldsymbol{q}''} \varepsilon^{-1}(\boldsymbol{q}, \boldsymbol{q}'') \, \varepsilon(\boldsymbol{q}'', \boldsymbol{q}') = \delta_{\boldsymbol{q},\boldsymbol{q}'}$.

From (2), $\boldsymbol{E}_{\text{tot}} = \boldsymbol{E}_{\text{ext}} + \boldsymbol{E}_{\text{ind}}$, and $\partial \boldsymbol{E}_{\text{ind}}(\boldsymbol{r}, t)/\partial \boldsymbol{r} = e n_{\text{ind}}(\boldsymbol{r}, t)/\varepsilon_0$ or $\boldsymbol{i q e E}_{\text{ind}}(\boldsymbol{q}, \omega) = 4\pi \in^2 n_{\text{ind}}(\boldsymbol{q}, \omega)$, it follows that for the matrix of electric polarizability, which is defined symbolically by $\varepsilon^{-1} = 1 + \sqrt{v} \, \alpha \, \sqrt{v}$ or in detail by

$$\varepsilon^{-1}(\boldsymbol{q}, \boldsymbol{q}'; \omega) = \delta_{\boldsymbol{q},\boldsymbol{q}'} + \sqrt{v_{\boldsymbol{q}}} \, \alpha(\boldsymbol{q}, \boldsymbol{q}'; \omega) \sqrt{v_{\boldsymbol{q}'}}, \quad v_{\boldsymbol{q}} = 4\pi \in^2/q^2, \tag{5}$$

the relation $n_{\text{ind}} = \alpha v_{\text{ext}}$ or

$$n_{\text{ind}}(\boldsymbol{q}, \omega) = \sum_{\boldsymbol{q}'} \alpha(\boldsymbol{q}, \boldsymbol{q}'; \omega) \, v_{\text{ext}}(\boldsymbol{q}', \omega) \tag{6}$$

holds ($v_{\text{ext}} = e u_{\text{ext}}$ is the mechanical potential acting on an electron), α describes the response of the system on the *unscreened* ("bare") external potential v_{ext}.

The electric susceptibility χ is (with respect to the information contained in it) equivalent to ε and α. For χ, it holds that

$$\varepsilon = 1 - \sqrt{v} \, \chi \sqrt{v} \quad \text{and} \quad n_{\text{ind}} = \chi v_{\text{tot}} \tag{6'}$$

(in symbols). Therefore, χ describes the response to the *screened* external potential $v_{\text{tot}} = v_{\text{ext}} + v_{\text{ind}}$.

For weak external perturbations, the material quantities ε and α are independent of the strength of the perturbative field and are determined by the metal considered only (linear response).

For the ground state, first-order time-dependent perturbation theory for n_{ind} and comparison with (6) yields

$$\alpha(\boldsymbol{q}, \boldsymbol{q}'; \omega) = \frac{1}{\hbar \Omega} \sum_n (\varrho_{\boldsymbol{q}}^\dagger)_{no}^* \, (\varrho_{\boldsymbol{q}'}^\dagger)_{no} \left[\frac{1}{\omega^+ - \omega_{no}} - \frac{1}{\omega^+ + \omega_{no}} \right] \tag{7}$$

according to Kubo. There, in (7) the sum is over the excited many-particle states of the system, $\hbar\omega_{no} = E_n - E_o$ are the excitation energies, $\omega^+ = \omega + i\delta$ with $\delta \gtrsim 0$, $\varrho_{\boldsymbol{q}} = \sum_i \exp(-\boldsymbol{q}\boldsymbol{r}_i)$ is the Fourier transform of the particle density operator $\varrho(\boldsymbol{r}) = \sum_i \delta(\boldsymbol{r} - \boldsymbol{r}_i)$ with $\varrho_{\boldsymbol{q}}^\dagger = \varrho_{-\boldsymbol{q}}$. In the Hartree approximation of independent particles, the many-particle states may be reduced to one-particle states, and (7) is simplified yielding (17).

The derivation of (7) is based on the concept that starting from the "unperturbed" system of interacting metal electrons (Hamiltonian H, eigenstates (E_n, Φ_n) from the ground state as initial state for $t \to -\infty$ (i.e. $\Psi(-\infty) = \Phi_0$), a state $\Psi(t)$ different from Φ_0 arises under the influence of an external perturbation $H^{\text{int}} = \int \mathrm{d}^3r\varrho(\boldsymbol{r})\, eu_{\text{ext}}(\boldsymbol{r}, t)$, this evolution being described within time-dependent perturbation theory. With $u_{\text{ext}}(\boldsymbol{r}, t) \sim \exp(\delta \cdot t)$, the perturbation is thought to be switched on quasistatically from 0 ($t = -\infty$) to 1 ($t = 0$). Forming the expectation value of $\varrho(\boldsymbol{r})$ with $\Psi(t)$, one obtains the perturbed particle density $n(\boldsymbol{r}, t)$. Thence with $n(\boldsymbol{r})$, the electron density in the unperturbed ground state Φ_0, the induced electron density $n_{\text{ind}}(\boldsymbol{r}, t) = n(\boldsymbol{r}, t) - n(\boldsymbol{r})$ results. The first perturbation order gives

$$n_{\text{ind}}(\boldsymbol{r}, t) = \frac{i}{\hbar} \int_{-\infty}^{t} \mathrm{d}t' \left(\Phi_0, [H^{\text{int}}(t'), \varrho(\boldsymbol{r}, t)]\, \Phi_0\right), \quad A(t) = \mathrm{e}^{\frac{i}{\hbar}Ht} A\, \mathrm{e}^{-\frac{i}{\hbar}Ht}.$$

Fourier transformation (the time-dependent exponentials yield the energy denominators) and comparison with (6) finally lead to (7), as required.

The lattice periodicity of the metal manifests itself in $\alpha(\boldsymbol{q}, \boldsymbol{q}'; \omega)$ and $\varepsilon(\boldsymbol{q}, \boldsymbol{q}'; \omega)$ being different from zero only in the case where $\boldsymbol{q}$ and $\boldsymbol{q}'$ differ from each other by a reciprocal lattice vector $\boldsymbol{G}$. One can express this by decomposing $\boldsymbol{q}$ and $\boldsymbol{q}'$ according to $\boldsymbol{q} = \boldsymbol{k} + \boldsymbol{G}$ and $\boldsymbol{q}' = \boldsymbol{k} + \boldsymbol{G}'$ into a part $\boldsymbol{k}$ that is located in the first Brillouin zone and a reciprocal lattice vector $\boldsymbol{G}$ or $\boldsymbol{G}'$, respectively. Then sums as in (2), (3), and (6) have to be carried out over $\boldsymbol{G}'$ only.

The terms arising through those elements of α that are non-diagonal with respect to $\boldsymbol{G}$ are called lattice local field corrections; for simple metals they can be taken into account perturbationally. In consequence of (5), ε^{-1} (and thus also ε) are known approximately if there is an approximation for α. Now, we ask which physical properties result from ε?

5.2.2 Loss Function and Reflectivity

For the probability per time, i.e. the transition probability γ for incident electrons (with about 10···50 keV) to lose energy $\hbar\omega$ and momentum $\hbar\boldsymbol{q}$ (with the system of metal electrons changing into one of its excited states), the golden rule of time-dependent perturbation theory (see for example Taut 83, p. 269) yields

$$\gamma_{\rm el}(\boldsymbol{q},\omega) = \frac{2}{\hbar\Omega}\frac{4\pi\epsilon^2}{q^2}\,[-\mathrm{Im}\,\varepsilon^{-1}(\boldsymbol{q},\boldsymbol{q};\omega)] \tag{8}$$

with the electrostatic interaction between the incident electrons and the metal electrons acting as the perturbation. Using (5), (8) may also be written as $\gamma_{\rm el}(\boldsymbol{q},\omega) \sim -\mathrm{Im}\, v_q \varkappa(\boldsymbol{q},\boldsymbol{q};\omega)\, v_q$. Considering the last v_q in this relation to be an external potential $v_{\rm ext}(\boldsymbol{q}',\omega) \sim v_q \delta_{\boldsymbol{q}',\boldsymbol{q}}$, with (6) it also holds that $\gamma_{\rm el}(\boldsymbol{q},\omega) \sim -\mathrm{Im}\, v_q n_{\rm ind}(\boldsymbol{q},\omega)$. In a plausible interpretation, this relation describes the interaction between the polarization cloud induced by the incident electron and the metal electron. After all, through this polarization cloud the incident electron experiences a reaction that retards its motion.

The quantity analogous to (8) for Compton scattering, which we discussed on an elementary level in [4.2.8], i.e. for the inelastic scattering of photons, is found as

$$\gamma_{\rm ph}(\boldsymbol{q},\omega) = \frac{2\pi\epsilon^2}{\hbar\Omega}\,\lambda\!\!\!^{-}{}_{\rm C}^2\,\frac{1+(\boldsymbol{e}\boldsymbol{e}')^2}{2KK'}\,q^2[-\mathrm{Im}\,\varepsilon^{-1}(\boldsymbol{q},\boldsymbol{q};\omega)] \tag{9}$$

with $K\boldsymbol{e}$ and $K'\boldsymbol{e}'$ the wave vectors of the photon before and after the interaction with the metal electrons, respectively, with $K'\boldsymbol{e}' = K\boldsymbol{e} - \boldsymbol{q}$ and $cK' = cK - \omega$; $\lambda\!\!\!^{-}{}_{\rm C} = \hbar/m\,|e|$ $(= 0.4 \times 10^{-12}\ \mathrm{m})$ is the Compton wavelength of the electrons. Thus $\hbar\omega$ proves to be the energy loss of the scattered light quantum in eqn. (9). Here the vector potential belonging to the photon (4.3-2) plays the role of the perturbing potential. It is this difference that also causes the prefactor to differ from that in eqn. (8).

For both electron and photon scattering, the quantity

$$P(\boldsymbol{q},\omega) = -\mathrm{Im}\,\varepsilon^{-1}(\boldsymbol{q},\boldsymbol{q};\omega), \tag{10}$$

to be referred to as the loss function, is thus relevant. In the first case it is multiplied by $1/q^2$, in the second, by q^2. Therefore, electron scattering experiments are precisely practicable only for small momentum changes $\hbar\boldsymbol{q}$ (i.e. small scattering angles), whereas photon scattering experiments are precise above all for large $\hbar\boldsymbol{q}$, both kinds of experiments thus mutually supplementing each other.

In both cases, a function

$$\varepsilon^{-1}(\boldsymbol{q},\omega) = \varepsilon^{-1}(\boldsymbol{q},\boldsymbol{q};\omega) \tag{11}$$

is required. However, while the knowledge of

$$P(\boldsymbol{q},\omega) = -\mathrm{Im}\,\varepsilon^{-1}(\boldsymbol{q},\omega) \tag{10'}$$

is necessary for calculating $\gamma_{\rm el}$ and $\gamma_{\rm ph}$, the reflection of light and electromagnetic waves of the adjoining ranges (from far infra-red with $\hbar\omega \approx 0.1$ eV up to far ultraviolet with $\hbar\omega \approx 50$ eV) results from $\varepsilon(\omega) = \varepsilon(0,\omega)$.

This fact is based on the following considerations (which incidentally, also make evident the physical content of $\varepsilon(\boldsymbol{q},\omega)$ as the macroscopic dielectric function in comparison to the microscopic dielectric matrix $\varepsilon(\boldsymbol{q},\boldsymbol{q}';\omega)$).

(1) Because the wavelength of the perturbation ($10^{-5}\dots10^{-8}$ m) as well as its penetration length (of the exponential decay) are large in comparison with the lattice constant (10^{-10} m), for the optical properties it is sufficient to consider macroscopic fields $\bar{E}_{\text{ind}}$ and $\bar{E}_{\text{tot}}$, which are averaged over a unit cell, instead of the microscopic fields E_{ind} and E_{tot}. The external perturbation is according to the supposition a macroscopic field: $\bar{E}_{\text{ext}} = E_{\text{ext}}$. As regards the Fourier expansion of the macroscopic quantities, this means that only Fourier components lying in the first Brillouin zone are different from zero:

$$\begin{aligned} \bar{E}_{\text{tot}}(\boldsymbol{k} + \boldsymbol{G}, \omega) &= E_{\text{tot}}(\boldsymbol{k}, \omega)\, \delta_{\boldsymbol{G},0}, \\ E_{\text{ext}}(\boldsymbol{k} + \boldsymbol{G}, \omega) &= E_{\text{ext}}(\boldsymbol{k}, \omega)\, \delta_{\boldsymbol{G},0}. \end{aligned} \tag{12}$$

Thus (2) yields

$$\bar{E}_{\text{tot}}(\boldsymbol{k}, \omega) = \varepsilon^{-1}(\boldsymbol{k}, \boldsymbol{k}; \omega)\, E_{\text{ext}}(\boldsymbol{k}, \omega) = \varepsilon^{-1}(\boldsymbol{k}, \omega)\, E_{\text{ext}}(\boldsymbol{k}, \omega), \tag{13}$$

with the macroscopic dielectric function $\varepsilon(\boldsymbol{k}, \omega) = [\varepsilon^{-1}(\boldsymbol{k}, \boldsymbol{k}; \omega)]^{-1}$ being introduced in the last step. This notation should be read as follows. Starting from $\varepsilon(\boldsymbol{k} + \boldsymbol{G}, \boldsymbol{k} + \boldsymbol{G}'; \omega) = \varepsilon_{\boldsymbol{G},\boldsymbol{G}'}(\boldsymbol{k}, \omega)$, invert the matrix $\varepsilon^{-1}_{\boldsymbol{G},\boldsymbol{G}'}(\boldsymbol{k}, \omega)$ with respect to $\boldsymbol{G}$, $\boldsymbol{G}'$ and take the number reciprocal to the matrix element corresponding to $\boldsymbol{G} = \boldsymbol{G}' = 0$. This means

$$\varepsilon^{-1}(\boldsymbol{k}, \omega) = [\varepsilon^{-1}_{\boldsymbol{G},\boldsymbol{G}'}(\boldsymbol{k}, \omega)]_{\boldsymbol{G}=\boldsymbol{G}'=0}. \tag{11'}$$

If off-diagonal elements, i.e. lattice local field corrections, may be neglected, $\varepsilon(\boldsymbol{k}, \omega) = \varepsilon(\boldsymbol{k}, \boldsymbol{k}; \omega)$ naturally holds; otherwise correction terms occur. The quantity $\varepsilon(\boldsymbol{q}, \omega)$ introduced in (11) thus generalizes $\varepsilon(\boldsymbol{k}, \omega)$ from $\boldsymbol{k}$ to $\boldsymbol{q} = \boldsymbol{k} + \boldsymbol{G}$.

(2) Since electromagnetic waves, as distinct from the quasistatic considerations of (1) to (4) with longitudinal field $\boldsymbol{E} \sim \boldsymbol{q}$, are characterized by transverse fields $\boldsymbol{E} \perp \boldsymbol{q}$, the question of whether the purely longitudinal components $\varepsilon(\boldsymbol{q}, \omega)$ of the dielectric tensor $\boldsymbol{\varepsilon}(\boldsymbol{q}, \omega)$ (which is defined in analogy to the macroscopic one introduced in (8)) are sufficient for the description of optical properties arises. The answer is yes. Using the assumption mentioned under (1) (wavelength $\gg$ atomic dimension) we see once more that it is sufficient to consider the material quantity $\boldsymbol{\varepsilon}(\boldsymbol{q}, \omega)$ for the simplifying limit $\boldsymbol{q} \to 0$ only. Now, in this limit the dielectric tensor $\boldsymbol{\varepsilon}(\omega) = \boldsymbol{\varepsilon}(0, \omega)$ may be obtained from its longitudinal components $\varepsilon_i(\omega) = \boldsymbol{e}_i \boldsymbol{\varepsilon}(0, \omega)\, \boldsymbol{e}_i$ in the directions of the main axis $\boldsymbol{e}_i$ ($i = 1, 2, 3$) belonging to the considered crystal symmetry:

$$\boldsymbol{\varepsilon}(\omega) = \sum_i \varepsilon_i \boldsymbol{e}_i \circ \boldsymbol{e}_i, \quad \varepsilon_i(\omega) = \boldsymbol{e}_i \boldsymbol{\varepsilon}(\omega)\, \boldsymbol{e}_i. \tag{14}$$

Such a representation certainly also holds for $\boldsymbol{q} \neq 0$, but the tensor components $\varepsilon_i(q\boldsymbol{e}, \omega) = \boldsymbol{e}_i \boldsymbol{\varepsilon}(q\boldsymbol{e}, \omega)\, \boldsymbol{e}_i$ occurring in this case do not prove to be longitudinal parts because of $\boldsymbol{e} \neq \boldsymbol{e}_i$. Only for $q = 0$ do the $\varepsilon_i(0, \omega)$ simultaneously have the meaning of the longitudinal parts in the direction of the corresponding main axis $\boldsymbol{e}_i$. This is because for $q = 0$ one cannot distinguish between longitudinal (plasmon-like) and transverse (photon-like) electromagnetic waves. In spite of the transversality of electromagnetic waves, the three longitudinal components of the dielectric tensor, $\varepsilon_i(\omega) = [\varepsilon_i^{-1}(0, 0; \omega)]^{-1}$, are thus sufficient for optical properties. For cubic (optically isotropic) metals, all these functions are equal to each other, for hexagonal (optically uni-axial) metals two functions are required: $\varepsilon_\parallel(\omega)$ and $\varepsilon_\perp(\omega)$.

Now, how do we obtain the optical properties from these macroscopic dielectric functions? For optically isotropic metals, from $\varepsilon(\omega) = \varepsilon_1(\omega) + i\varepsilon_2(\omega)$ the complex refraction index $n(\omega) = n_1(\omega) + in_2(\omega)$ results according to

$$n^2 = \varepsilon \curvearrowright \begin{cases} n_1^2 = \dfrac{1}{2}\left(\varepsilon_1 + \sqrt{\varepsilon_1^2 + \varepsilon_2^2}\right) \\[2ex] n_2^2 = \dfrac{1}{2}\left(-\varepsilon_1 + \sqrt{\varepsilon_1^2 + \varepsilon_2^2}\right). \end{cases} \tag{15}$$

$n(\omega)$, in turn, determines the reflectivity $R(\omega)$ for perpendicular incidence according to

$$R = \left|\frac{1 - n}{1 + n}\right|^2 = \frac{(1 - n_1)^2 + n_2^2}{(1 + n_1)^2 + n_2^2}. \tag{16}$$

For non-ferromagnetic metals, (15) and (16) follow from the Maxwell equations in connection with the definition of $n(\omega)$ according to $k^2 = (\omega/c)^2\, n(\omega)$ and of $R(\omega)$ as the ratio of the incident and the reflected energy current densities.

For optically uniaxial metals, there are quantities $n_\parallel$ and $n_\perp$ as well as $R_\parallel$ and $R_\perp$ corresponding to $\varepsilon_\parallel$ and $\varepsilon_\perp$. To determine these two reflectivities, the surface is prepared in such a way that it lies parallel to the optical axis ($\sim \boldsymbol{e}_c$). The polarization can then be applied both parallel to the optical axis (in accordance with $\boldsymbol{E} \parallel \boldsymbol{e}_c$, thus $\varepsilon_\parallel$, $n_\parallel$, $R_\parallel$) and perpendicular to it ($\boldsymbol{E} \perp \boldsymbol{e}_c$, thus $\varepsilon_\perp$, $n_\perp$, $R_\perp$).

5.2.3 Approximations for $\varepsilon(\boldsymbol{q}, \omega)$

Our former considerations may be summarized in the following scheme:

$$\left.\begin{matrix} E_n \\ \Phi_n \end{matrix}\right\} \xrightarrow{(7)} \alpha(\boldsymbol{q}, \boldsymbol{q}; \omega) \xrightarrow{(5)} \varepsilon^{-1}(\boldsymbol{q}, \boldsymbol{q}; \omega)$$

$$\xrightarrow{(11)} \varepsilon(\boldsymbol{q}, \omega) - \begin{cases} \xrightarrow{(10')} P(\boldsymbol{q}, \omega) - \begin{cases} \xrightarrow{(8)} \gamma_{\mathrm{el}}(\boldsymbol{q}, \omega) \\ \xrightarrow{(9)} \gamma_{\mathrm{ph}}(\boldsymbol{q}, \omega) \end{cases} \\ \xrightarrow{(15)} n(\omega) \xrightarrow{(16)} R(\omega). \end{cases}$$

The first step contains the solution of the many-particle problem of the Bloch electrons.

The simplest approximation that contains both the one-particle (Bloch electrons as particle-like excitations or quasiparticles) and the many-particle (plasmons as collective excitations) aspects is the so-called random phase approximation (RPA, see for example Ziesche, Lehmann et al. 83, pp. 36, 261):

$$\chi^0(\boldsymbol{q}, \boldsymbol{q}'; \omega) = \frac{1}{\hbar\Omega} \sum_{k',k''} (\varrho_{\boldsymbol{q}}^\dagger)^*_{k''k'} (\varrho_{\boldsymbol{q}'}^\dagger)_{k''k'} \frac{f^0_{k''} - f^0_{k'}}{\omega_{k''k'} - \omega^+}, \tag{17}$$

$$\alpha_{\mathrm{RPA}} = \chi^0(1 - v\chi^0)^{-1}, \tag{18}$$

$$\varepsilon_{\mathrm{RPA}} = 1 - \sqrt{v}\, \chi^0 \sqrt{v}, \quad \varepsilon_{\mathrm{RPA}}^{-1} = 1 + \sqrt{v}\, \alpha_{\mathrm{RPA}} \sqrt{v}. \tag{19a, b}$$

The matrix elements occurring in (17) should be constructed with Bloch wave functions $\psi_k(\boldsymbol{r}, \sigma) = \delta_{s,\sigma}\psi_{\boldsymbol{k}\nu}(\boldsymbol{r})$. $k \triangleq (\boldsymbol{k}, \nu, s)$ thus summarizes the Bloch vector $\boldsymbol{k}$, the band index ν, and the spin projection s. Furthermore $\hbar\omega_{k''k'} = \varepsilon_{k''} - \varepsilon_{k'}$, and f^0_k is the Fermi distribution function for $T = 0$. In (18) and (19), the short-hand notation introduced in (5) is used.

For a deeper understanding of RPA, the following is true: (6) and (6′) introduce material quantities according to $n_{\mathrm{ind}} = \alpha v_{\mathrm{ext}}$ and $n_{\mathrm{ind}} = \chi v_{\mathrm{tot}}$ with $v_{\mathrm{tot}} = v_{\mathrm{ext}} + v_{\mathrm{ind}}$ and $v_{\mathrm{ind}} = v n_{\mathrm{ind}}$ (the latter being a consequence of the Poisson equation). (7) gives a rigorous representation of α. Now, if one wants to evaluate this representation in the one-particle approximation of non-interacting Bloch electrons — comparing (7) with (17) shows that this would mean $\alpha \approx \chi^0$ — the peculiarities of Coulomb interaction (long range, screening), would produce difficulties, because according to $n_{\mathrm{ind}} = \alpha v_{\mathrm{ext}}$ one would describe the response of the system to the bare external perturbation only. The actually existing polarization of the medium caused by the external perturbation, i.e. the screening of the external perturbation would thus be neglected. The essence of the RPA consists of considering not α (describing the reaction to the bare external perturbation) but χ (describing the reaction to the screened external perturbation) within the one-particle approximation $\chi \approx \chi^0$. Since $n_{\mathrm{ind}} \approx \chi^0 v_{\mathrm{tot}}$, $n_{\mathrm{ind}} \approx \chi^0(v_{\mathrm{ext}} + v n_{\mathrm{ind}})$, a self-consistent treatment of the screening in the simplest approximation according to the Hartree method results from successive iteration (perturbation expansion) or directly by solving the equation with respect to n_{ind}:

$$n_{\mathrm{ind}} \approx (1 - \chi^0 v)^{-1} \chi^0 v_{\mathrm{ext}} = \chi^0(1 - v\chi^0)^{-1} v_{\mathrm{ext}}.$$

Comparison with $n_{\mathrm{ind}} = \alpha v_{\mathrm{ext}}$ indeed gives (18). Thus the RPA describes the response of non-interacting Bloch electrons to the self-consistently screened external perturbation. Writing (18) in the form $\alpha_{\mathrm{RPA}} = \chi^0 + \chi^0 v \chi^0 + \cdots$, one recognizes that RPA proves to be a certain partial summation starting from the rigorous expression (7) and its perturbational expansion into powers of the Coulomb interaction, the polarization effects being taken into account through electron-electron interaction in the simplest approximation.

Evaluating (17) and (19) with free particles, $\psi^0_{\boldsymbol{k}}(\boldsymbol{r}) = (1/\sqrt{\Omega}) \exp(i\boldsymbol{k}\boldsymbol{r})$, $\varepsilon^0_{\boldsymbol{k}} = \hbar^2\boldsymbol{k}^2/2m$, we get some idea of the properties of $\varepsilon_{\mathrm{RPA}}(\boldsymbol{q}, \boldsymbol{q}'; \omega)$. The non-diagonal elements vanish in this case and for

$$\varepsilon^0_{\mathrm{RPA}}(\boldsymbol{q}, \boldsymbol{q}; \omega) = 1 - v_q\chi^0(\boldsymbol{q}, \omega) = \varepsilon^0(\boldsymbol{q}, \omega)$$

the Lindhard function results. For $\omega = 0$ or $\boldsymbol{q} = 0$ one obtains

$$\varepsilon^0(\boldsymbol{q}) = \varepsilon^0(\boldsymbol{q}, 0) = 1 + \frac{q_{\mathrm{TF}}^2}{q^2}\, l\left(\frac{q}{2k_{\mathrm{F}}}\right),$$

$$l(x) = \frac{1}{2}\left[1 + \frac{1 - x^2}{2x} \ln\left|\frac{1 + x}{1 - x}\right|\right], \quad q_{\mathrm{TF}}^2 = \frac{4k_{\mathrm{F}}}{\pi a_0}, \tag{20a}$$

$$\varepsilon^0(\omega) = \varepsilon^0(0, \omega) = 1 - \frac{\omega_{\mathrm{pl}}^2}{\omega^2} - i\pi\omega_{\mathrm{pl}}^2 \frac{\mathrm{d}}{\mathrm{d}\omega}\, \delta(\omega),$$

$$\omega_{\mathrm{pl}}^2 = \frac{4\pi e^2 n}{m} \tag{20b}$$

respectively, with ω_{pl} the classical plasma frequency, and $1/q_{\mathrm{TF}}$ the Thomas-Fermi screening length. Altogether, the trend depicted in fig. 5.2-1 for $\varepsilon_1^0(\boldsymbol{q}, \omega) = \mathrm{Re}\, \varepsilon^0(\boldsymbol{q}, \omega)$ and $\varepsilon_2^0(\boldsymbol{q}, \omega) = \mathrm{Im}\, \varepsilon^0(\boldsymbol{q}, \omega)$ results; the singular behaviour of $\varepsilon_2^0(\boldsymbol{q}, \omega)$ for $\boldsymbol{q} = 0$ has not been recorded in the figure.

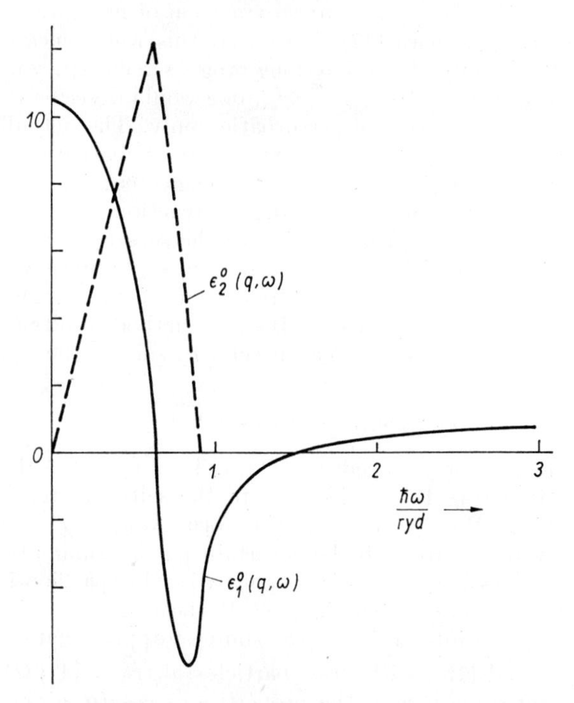

Fig. 5.2-1. Dielectric function within RPA for free electrons with the valence electron density of Be and $q = 0.4 \cdot 2\pi/c$ (c = lattice spacing along the hexagonal axis)

For some purposes (lattice structure and dynamics) the RPA is not sufficient, because RPA means that according to $\chi \approx \chi^0$ and thus $n_{\text{ind}} \approx \chi^0 v_{\text{tot}}$ only non-interacting electrons respond to the total potential $v_{\text{tot}} = v_{\text{ext}} + v_{\text{ind}}$. Here we have not taken into consideration the fact that the density of the remaining electrons is diminished locally at the position of a certain electron as a consequence of exchange and correlation (local field). This local density decrease means that the potential $v_{\text{ind}}^{(\text{e})}$ an electron experiences is smaller than the potential $v_{\text{ind}} = vn_{\text{ind}}$ a weak test charge would experience, i.e., $v_{\text{ind}}^{(\text{e})} = (1 - G)\, v_{\text{ind}} = (1 - G)\, vn_{\text{ind}}$, where G is the so-called local field correction (LFC). (This LFC due to correlation is for the homogeneous electron gas a function $G(\boldsymbol{q}, \omega)$. Often the frequency dependence is not taken into account and only static LFC's $G(\boldsymbol{q})$ are used.) For the total potential acting on an electron, $v_{\text{tot}}^{(\text{e})} = v_{\text{ext}} + v_{\text{ind}}^{(\text{e})} = v_{\text{tot}} - Gvn_{\text{ind}}$ results. As within the RPA, non-interacting electrons respond to this potential according to $n_{\text{ind}} \approx \chi^0 v_{\text{tot}}^{(\text{e})}$. Inserting $v_{\text{tot}}^{(\text{e})}$, resolving with respect to n_{ind}, and comparing to $n_{\text{ind}} = \chi v_{\text{tot}}$ yields

$$\chi(\boldsymbol{q}, \omega) = \frac{\chi^0(\boldsymbol{q}, \omega)}{1 + G(\boldsymbol{q})\, v_q \chi^0(\boldsymbol{q}, \omega)}, \quad v_q = \frac{4\pi e^2}{q^2}. \tag{21}$$

For $G(\boldsymbol{q}) \equiv 0$, one obtains the RPA again with $\chi = \chi^0$, and vice versa one can assume the local field corrections to be produced by a partial summation that goes beyond the RPA. Inserting (21) into (6′), one obtains

$$\varepsilon(\boldsymbol{q}, \omega) = 1 - v_q \chi(\boldsymbol{q}, \omega) = 1 - \frac{v_q \chi^0(\boldsymbol{q}, \omega)}{1 + G(\boldsymbol{q})\, v_q \chi^0(\boldsymbol{q}, \omega)}, \tag{21′}$$

which is reminiscent of the LFC of the Clausius-Mosotti or the Lorentz-Lorenz formulae (see, for example, Lenk 76). As distinct from the lattice LFC's discussed in [5.2.2] the corrections caused by exchange and correlation are referred to as correlation LFC's. While $\varepsilon(\boldsymbol{q}, \omega)$, according to $v_{\text{tot}} = v_{\text{ext}}/\varepsilon$, describes the screening of an external perturbation as experienced by a weak test charge, which does not perturb the electron gas, another dielectric function $\varepsilon^{(\text{e})}(\boldsymbol{q}, \omega)$ according to $v_{\text{tot}}^{(\text{e})} = v_{\text{ext}}/\varepsilon^{(\text{e})}$ describes the screening of an electron that belongs to the system and thus polarizes its surroundings (local field.) We have

$$\varepsilon^{(\text{e})}(\boldsymbol{q}, \omega) = 1 - v_q[1 - G(\boldsymbol{q})]\, \chi^0(\boldsymbol{q}, \omega). \tag{22}$$

From a comparison of (21) with (22), $\varepsilon^{(\text{e})}/\varepsilon = \chi^0/\chi$ follows. While ε occurs in the theory of phonons and the electron-phonon interaction, $\varepsilon^{(\text{e})}$ for instance determines the pseudopotential of ions acting on the conduction electrons of simple metals and which are self-screened. The function given by Singwi et al. (70) has been widely used for $G(\boldsymbol{q})$. Recently, besides these statical (only $\boldsymbol{q}$-dependent) LFC's, dynamical ($\boldsymbol{q}$-and ω-dependent) LFC's have also received attention. For further details, see Ziesche, Lehmann et al. 83, p. 45, for example.

5.2.4 Results for $P(\boldsymbol{q}, \omega)$, $n(\omega)$ and $R(\omega)$

From $\varepsilon^0(\boldsymbol{q}, \omega)$, the loss function $P^0(\boldsymbol{q}, \omega)$ results according to (10′). It is different from zero only in the hatched area in Fig. 5.2-2 (where $\varepsilon_2^0(\boldsymbol{q}, \omega) \neq 0$) and on the curve $\omega_{\rm pl}(q)$ following from $\varepsilon^0(\boldsymbol{q}, \omega) = 0$ (where both $\varepsilon_1^0(\boldsymbol{q}, \omega) = 0$ and $\varepsilon_2^0(\boldsymbol{q}, \omega) = 0$). These different branches correspond to different excitations of valence electrons. Transferred energies and momenta lying in the hatched area in Fig. 5.2-2 are accepted by a single electron. This electron goes over to an (unoccupied) state outside the Fermi sphere, thus leaving behind a hole in the Fermi sphere: a particle-hole pair is generated. For its possible energies and momenta one just finds the area of single-pair excitations in Fig. 5.2-2 according to the Pauli principle.

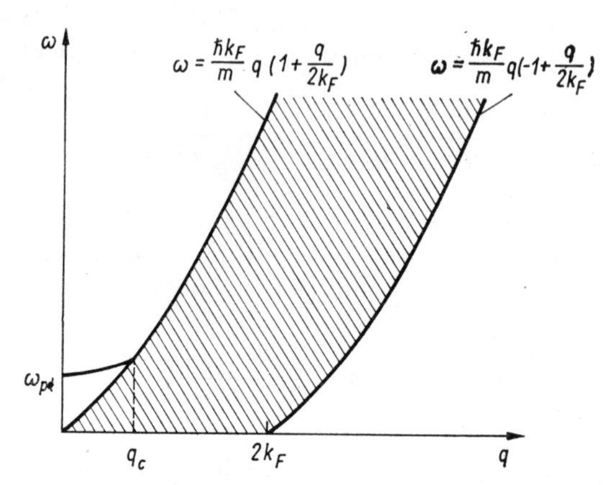

Fig. 5.2-2. Spectrum of individual and collective excitations of the homogeneous electron gas within RPA; single-pair excitations in the hatched stripe and undamped plasmons along the curve starting at $q = 0$, $\omega = \omega_{\rm pl}$

Furthermore, $P^0(\boldsymbol{q}, \omega)$ has a δ-like peak along the curve $\omega_{\rm pl}(q)$ that begins at $q = 0$ with the classical plasma frequency $\omega_{\rm pl}$ and dips into the one-pair continuum at q_c. For such energy and momentum transfers, the system of valence electrons is excited as a whole. These collective excitations are called plasmons, and $\omega_{\rm pl}(q)$ is their dispersion relation.

They are undamped within the approximation of free electrons and RPA which is discussed first. However, in reality, deviations from free-electron behaviour (band structure effects) and multi-pair excitations (effects of electron correlation going beyond RPA, described by the correlation LFC's $G(\boldsymbol{q})$) give rise to a finite plasmon damping. Instead of a δ-function with vanishing width we have a δ_ε-function with a finite width $2\Gamma(q)$, which is still increased by the additional influence of lattice defects, phonons, and the finite lifetime of the Bloch states.

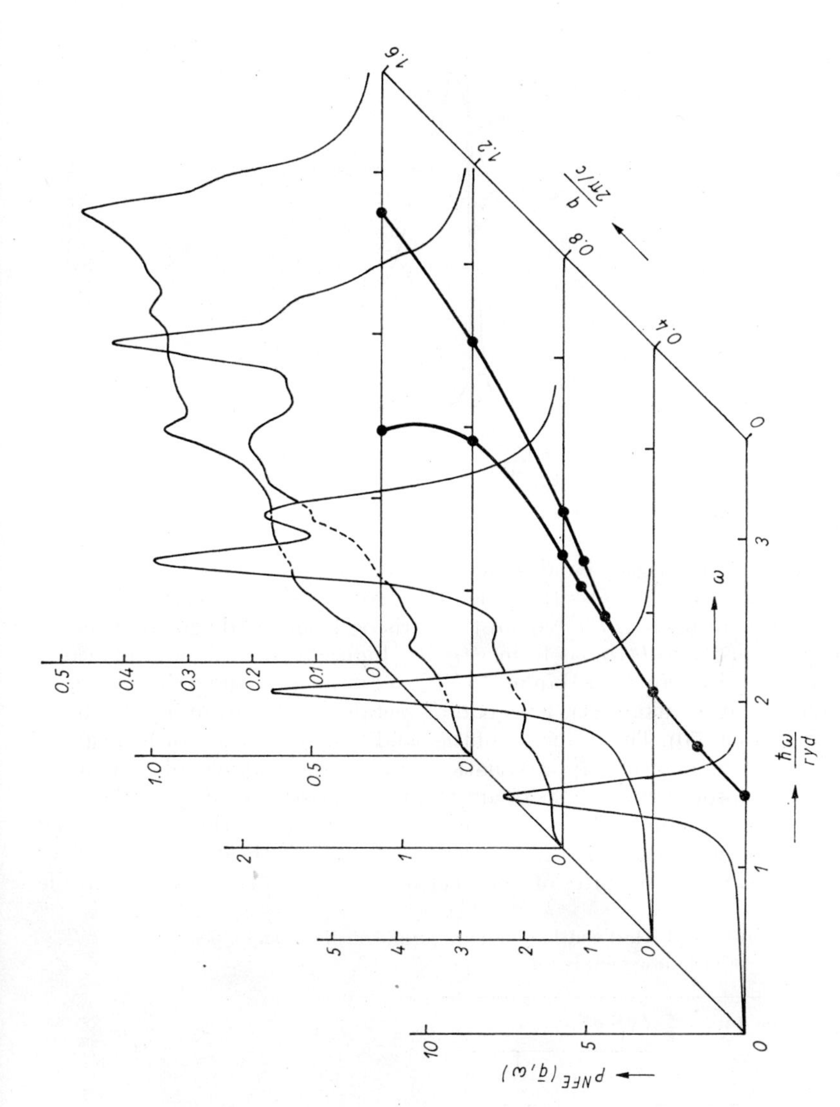

Fig. 5.2-3. a) Calculated loss function of Be. The double peak structure for $q \gtrsim q_c$ is also observed experimentally. The value of q_c is about $0.7 \cdot 2\pi/c$

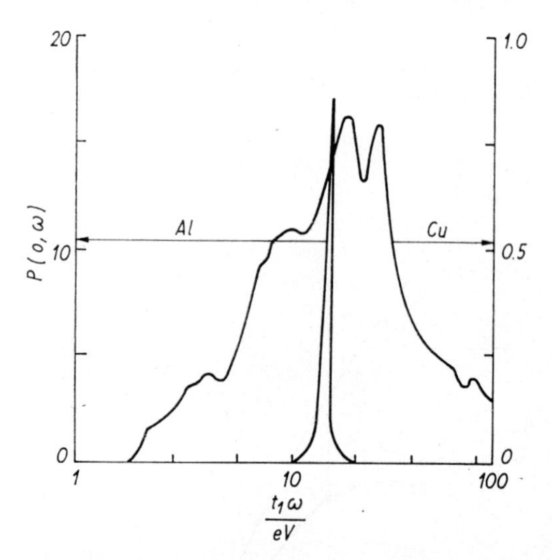

Fig. 5.2-3. b) Experimental loss functions of Al and Cu for $q \to 0$

It is useful to investigate band structure effects for simple metals first. Here perturbation treatment shows that the off-diagonal elements of the dielectric matrix may be neglected, $P^{\mathrm{NFE}}(\boldsymbol{q}, \omega)$ thus arising from the diagonal elements $\varepsilon_{\mathrm{RPA}}(\boldsymbol{q}, \boldsymbol{q}; \omega)$ just like $P^0(\boldsymbol{q}, \omega)$ from $\varepsilon^0(\boldsymbol{q}, \omega)$. Figure 5.2-3a) shows the results for Be, as given by Taut and Hanke (76), fig. 5.2-3b) shows in addition $P(0, \omega)$ for Al and Cu after Lukas (81). For alkaline metals, $\Gamma(0)$ is given in table 5.2-1 following Paasch (70). The influence of the band structure must not be neglected in the high-momentum approximation either, which above all is required for inelastic photon scattering (see Taut 83). In this case $\varepsilon_2(\boldsymbol{q}, \omega) \ll \varepsilon_1(\boldsymbol{q}, \omega) \approx 1$ holds, and (10) is simplified to $P(\boldsymbol{q}, \omega) \approx \varepsilon_2(\boldsymbol{q}, \omega)$. Indeed those unoccupied Bloch states occurring in (17) that are located far above the Fermi level, may be well approximated by states of free electrons $\psi^0_{\boldsymbol{k}}$, $\varepsilon^0_{\boldsymbol{k}}$, but the Compton profile

Table 5.2-1. Half width at half maximum of the plasmon peak of the alkaline metals for $q = 0$

metal	$\Gamma(0)$/eV	
	theor.	exper.
Li	2.55	2.5
Na	0.1	0.4
K	0.25	0.26
Rb	0.64	0.6
Cs	0.96	1.2

is given by

$$I(\varepsilon, \varepsilon', \boldsymbol{p}, s) \sim \varepsilon \sum_{\mathrm{i(f)}} |\langle\psi_\mathrm{i}| H^I |\psi_f\rangle|^2 \, \delta(\varepsilon + \varepsilon_\mathrm{i} - \varepsilon_\mathrm{f} - \varepsilon'). \tag{1}$$

It is measured as a function of excitation energy ε and depends also on energy ε', momentum $\boldsymbol{p}$, and spin s of the emitted quantum. The right-hand side contains the square of the matrix element of the interaction H^I and describes the probability of the transition from the initial state ψ_i to the final state ψ_f of the system. In these states, all internal (quasi) and external (test) particles participating in the interaction process have to be included. The δ-function ensures the energy balance. If the change of momentum (angular dependence) can be determined by the equipment, then a δ-function for the momentum balance has to be added. Furthermore, one has to average or sum over the possible final (f) and initial (i) states, respectively, according to the particular experimental conditions.

If one of both participating states is known exactly (e.g. the low-lying initial state ψ_i) from I, conclusions may be drawn on the other state (ψ_f). These conclusions concern both the energy ε_f and the number of states available at this energy (i.e. the DOS $D(\varepsilon)$), respectively, and also via the amount of matrix element information on the shape of the wave function. A broadening of the levels as a consequence of their finite life time and of other experimental conditions reduces the accuracy of the information available.

For transitions of type (3) which may be induced, for example, by optical excitation, the energy difference between initial and final state is given, and both vary themselves across the whole band. As a result, so-called mixed DOS are measured containing coupled information on both initial and final states.

4.3.2 X-Ray Emission Spectra

In the following, the emission of X-rays will be discussed in more detail. This corresponds to processes (4) to (6). In (1), the interaction $H^\mathrm{I} = |e|\, \boldsymbol{pA}/m$ with the radiation field is determined by the vector potential $\boldsymbol{A} = \sqrt{\hbar/2\varepsilon_0\Omega\omega}\,\boldsymbol{n} \exp(i\boldsymbol{kr})$ with the polarization vector $\boldsymbol{n}$. As the extension of the tightly bound empty state is small in comparison with the wave length of a soft X-ray quantum ($ka \approx 10^{-1} \cdots 10^{-2}$), we can use for H^I the so-called dipole approximation:

$$H^\mathrm{I} = \frac{|e|}{m} \boldsymbol{pA} \sim \omega^{-1/2} \boldsymbol{pn}\, \mathrm{e}^{i\boldsymbol{kr}} \approx \boldsymbol{np}\omega^{-1/2}. \tag{2}$$

Since we consider a spontaneous emission process in (1), the dependence on the energy ε of the radiation incoming on the sample vanishes. We denote the energy of the emitted photon by $\varepsilon' = \hbar\omega$. The matrix element of momentum is replaced by that of the position operator[1]):

$$\langle\psi_\mathrm{i}\, |\boldsymbol{np}|\, \psi_\mathrm{f}\rangle = im\omega\langle\psi_\mathrm{i}\, |\boldsymbol{nr}|\, \psi_\mathrm{f}\rangle. \tag{3}$$

[1]) This relation is a consequence of the commutation relation $[H, \boldsymbol{r}] = \hbar\boldsymbol{p}/im$ and of the fact that ψ_i and ψ_f are eigenstates of H with energies ε_i and ε_f, whereby $\hbar\omega = \varepsilon_\mathrm{i} - \varepsilon_\mathrm{f}$ holds.

If in (3) the matrix elements of the operator $\boldsymbol{r}$ do not depend on the energy, then the fine-structure of the X-ray line according to (1) to (3) gives direct information on the density of occupied states,

$$I(\omega) \sim \omega^2 D(\omega + \varepsilon_f/\hbar). \tag{4}$$

However, the exact structure will be influenced by the energy dependence of the matrix elements. For simple metals, the matrix elements of the position operator calculated by means of pseudo wave functions agree well with the experiment (better than the matrix elements of the momentum operator). For the spectra of transition metals, the selection rules concerning the angular momenta of initial (l) and final (l') state,

$$l' = l \pm 1 \tag{5}$$

contained in (3) play a decisive role. According to the atomic spectra, the solid spectra are denoted after the shells for the final states of the transition, K, L, M, etc. These tightly bound final states have a well defined angular momentum l'. Because of (5), the contributions of the conduction states belonging to l have to be selected according to the representation of the Bloch function

$$\psi_{\boldsymbol{k}}(\boldsymbol{r}) = \sum_L i^l c_L(\boldsymbol{k})\, R_l(r, \varkappa)\, Y_L(\boldsymbol{r}) \tag{6}$$

inside the Wigner-Seitz cell. Comparison with experiment now yields, instead of (4), the "partial" density of states (m = magnetic quantum number)

$$D_l(\varepsilon) = \sum_{\boldsymbol{k},m} |c_L(\boldsymbol{k})|^2\, \delta(\varepsilon - \varepsilon_{\boldsymbol{k}}). \tag{7}$$

The energy dependence of the residual integrals via the radial wave functions may be neglected in most cases for transition metals.

Using polarized X-rays and single crystals makes available information on the orientation of the wave function via the direction dependence of the matrix elements (3).

For heavy transition elements ($Z > 40$), the spin-orbit interaction of the states has a strong influence on the shape of the spectra (Nemoshkalenko et al. 82). The split of the d-states as a consequence of this interaction has the same order as the width of the d-band. For the electronic structure of actinides, see Weinberger (84), (90).

As the low-lying states of a certain type i of atom are well coordinated to their energy, the energy dependent matrix element tests the widely extended states of the conduction band precisely at the site of atom $\boldsymbol{R}_i$, since the wave function of the final state $\psi_f(\boldsymbol{r} - \boldsymbol{R}_i)$ is well localized at that site (partial "local" DOS). Of course, very low-lying states are less informative because of large level broadening as a consequence of their reduced life time.

Interpreting the experimental spectra, many-particle effects (e.g. plasmon excitations) have also to be taken into account. An important problem arises from the question of the extent to which the Coulomb interaction of the hole state influences the one-electron function of the Bloch state. Recently, it was shown that the final state of the process without low-lying hole states plays the decisive role: consequently, the matrix element may be calculated using ordinary Bloch wave functions (von Barth, Grossmann 79, Mahan 80). However, a strict proof on the basis of many-particle theory of strongly bound hole states is still an open question.

4.3.3. Photo-Electron Spectroscopy

With the availability of tunable high-power radiation sources such as synchrotron radiation sources, photo-electron spectroscopy (PES), in the ultraviolet (UPS), and in the X-ray spectral region (XPS) have given increasing importance to the experimental investigation of the electronic structure of solids. Thereby, the incoming light quantum with energy $\varepsilon = \hbar\omega$ excites, corresponding to processes (1) and (2) in fig. 4.3-1, an electron which will be transported in a second step to the surface of the sample and will escape from the solid in a third step, undergoing reflection and scattering effects at the surface. As a consequence of the strongly inelastic processes during the second step with UPS, only electrons from the uppermost layers may be analysed, whereas for XPS with electron energies of about 1 keV with an electronic mean free path of the order of 5 to 10 nm bulk effects are certainly seen. The dipole approximation (2) is also applicable for the excitation process.

For UPS, the change in momentum of the electrons is small because of the small momentum of the incident photon. Therefore, we obtain the "perpendicular" transitions represented in fig. 4.3-2. Generally speaking, the inter-

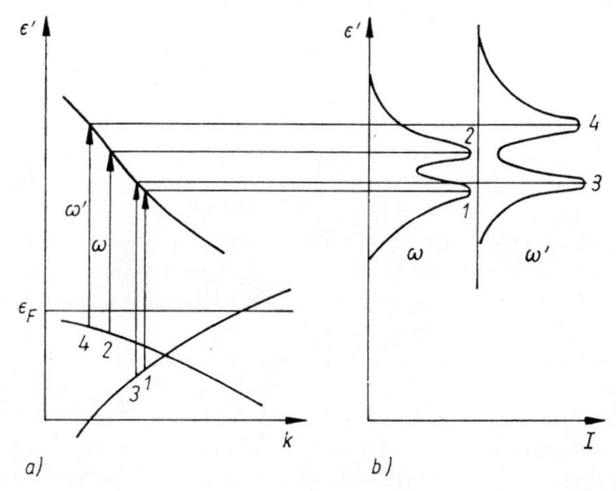

Fig. 4.3-2. Explanation of the kind of information obtainable by emission of photoelectrons

pretation of the spectra is the more complicated, the more unknown states are involved in the transition probabilities. Advantages arise for PES because the external conditions (such as the energy of the incoming radiation) may be varied widely as shown schematically in fig. 4.3-2. In fig. a) a band structure with two crossing occupied bands is seen.

Incoming photons with frequency ω induce transitions (1) and (2). This causes the shape of the intensity as a function of energy ε' of the emitted electron as represented in fig. 4.3-2b). The transitions (1) and (2) are broadened to the two neighbouring maxima (1 and 2). For a somewhat increased incoming energy $\hbar\omega' > \hbar\omega$, transitions (3) and (4) result. As a consequence of the opposed tendencies of both crossing bands, the distance between the maxima (3) and (4) of the intensity distribution function (for the incoming energy $\hbar\omega'$) changes strongly in comparison with the change $\hbar(\omega - \omega')$. Therefore, by a change of ω a variety of valuable information may be gained. The interpretation is facilitated if the position of a "reference band" is known. This is the case if dispersionless surface states are present. In the angle-resolved PES (ARPES), the direction of the $\boldsymbol{k}$-vector is given by the direction of the observed emitted electron. Therefore, it is thus possible to measure directly the dispersion relation of occupied and unoccupied bands (Eastman, Himpsel 81). From the height of the intensity maxima, information on the matrix elements may be gained. Using polarized radiation enables determination of symmetry properties of the bands. The analysis of the spins of the emitted electrons allows assertions on spin-polarized bands in magnetically ordered systems (ferromagnetic exchange splitting in Co: Eastman et al. 80). These measurements give indications that the exchange splitting evaluated with the potential V_{xc} (1.3-13) of von Barth and Hedin (72) becomes too large.

In the XPS, the energies of the excited electrons are so large that the final state may be described by nearly free electrons. On the other hand, the momentum change is of the order of the diameter of the Brillouin zone and, therefore, cannot be neglected. The diffraction of the emitted electron at the surface of the sample complicates the treatment of energy and momentum balance. Because of the large momenta of the emitted electrons, the angular resolution of the spectrometer limited to several degrees induces uncertainties in the components of momenta perpendicular to the observation direction of the order of the Brillouin zone diameter. Therefore, the observed intensity gives no information on the dispersion relation but only on partial densities of states. Moreover, it is of advantage that the density of final states at high energies changes only weakly and monotonously.

Besides statements on band structure, the XPS also provides information on charge relations and chemical bonds in the solid investigated by using the shifts in absolute line position compared with free atoms and known standard materials. The absolute position of the lines characterizes as do characteristic X-rays, the chemical elements in which the initial state is localized. As a consequence, the PES is often also denoted ESCA (Electronic Spectroscopy for Chemical Analysis) and used for chemical analysis (Siegbahn 67).

The intensity of the XPS photoelectrons from core levels show a strong angular dependence (ARXPS). It is caused by the scattering of the photoelectron at the atoms in the neighbourhood of the emitter. The evaluation of the spectra gives information on the lattice spacing, the surface reconstruction, the position of adsorbate atom etc. — To explain ARXPS it is necessary to go beyond the plane wave approximation for the final state. It is a sum of spherical waves with an origin at the emitter and at the neighbouring atoms. The single scattering process can be described by the plane wave approximation, but more refined considerations include curved wave and multiple scattering corrections (Rennert and Hung 88).

5 Further Electronic Properties

5.1 Density of States, Heat Capacity, Transition Temperature of Superconductivity, Magnetic Susceptibility, and Transport Quantities

The DOS at the Fermi energy, D_F, essentially determines the electronic heat capacity. The electron-phonon interaction modifies this relation, causes superconductivity, and determines its transition temperature. D_F and the electron-electron interaction determine the susceptibility of the Pauli spin paramagnetism. Transport quantities such as electrical resistivity can be calculated by solving the Boltzmann equation; the latter contains transition probabilities, which are determined by the band structure and the phonon spectrum, respectively.

5.1.1 Density of States and Heat Capacity

The DOS $D(\varepsilon)$ follows from the band structure $\varepsilon_{\boldsymbol{k},\nu}$ as the number of electron states per energy interval and unit cell according to (cf. 1.2-24 as well)

$$D(\varepsilon) = \frac{2}{N} \sum_{\substack{\boldsymbol{k} \in \mathrm{BZ} \\ \nu}} \delta(\varepsilon - \varepsilon_{\boldsymbol{k},\nu}) = \frac{2\Omega_0}{(2\pi)^3} \sum_{\nu} \int_{\mathrm{BZ}} \mathrm{d}^3k \delta(\varepsilon - \varepsilon_{\boldsymbol{k},\nu}) . \tag{1}$$

The factor 2 takes into account the two spin directions. Each of the N unit cells has a volume Ω_0. The Fermi energy is determined by the requirement that the integral over $D(\varepsilon)$ with ε_F as the upper bound just equals the total number of electrons per unit cell.

The partial information about the band structure contained in $D(\varepsilon)$ determines the equilibrium thermodynamics of the electron system via

$$U_{\mathrm{el}}(T) = N \int \mathrm{d}\varepsilon D(\varepsilon)\, f(\varepsilon, T)\, \varepsilon , \tag{2}$$

which contributes to the internal energy $U(T) = U_{\mathrm{el}}(T) + U_{\mathrm{ph}}(T)$. $U_{\mathrm{el}}(T)$ contains with $f(\varepsilon, T)$ the Fermi-Dirac distribution function (1.1-7). The contribution of the phonon system, $U_{\mathrm{ph}}(T)$, being analogous to (2), contains the Bose-Einstein distribution function (5.3-13) according to (5.3-14). As for $U(T)$, the heat capacity at constant volume Ω

$$C_\Omega(T) = \left(\partial U(T)/\partial T\right)_\Omega \tag{3}$$

also consists of two parts, $C_\Omega(T) = C_{\mathrm{el}}(T) + C_{\mathrm{ph}}(T)$. At low temperatures T ($\ll \theta_D$, the Debye temperature), it holds that $C_{\mathrm{el}}(T) \sim T$ and $C_{\mathrm{ph}}(T) \sim T^3$, and thus (for non-magnetic metals)

$$C_\Omega(T) = \gamma T + AT^3 + \cdots. \tag{4}$$

The term proportional to T is caused by the smearing out of the Fermi edge. It follows from the evaluation of $\partial U_{el}/\partial T$ with $U_{el}(T)$ from (2) that at low T, $\partial f(\varepsilon, T)/\partial T$ essentially differs from zero only in a small interval $\sim k_B T$ around the Fermi energy ε_F (see fig. 1.1-1). There $D(\varepsilon)$ can be considered constant and can be replaced by a constant factor $D_F = D(\varepsilon_F)$, which simplifies the evaluation of the integral: $\gamma \sim D_F$. The evaluation of the remaining integral yields the proportionality factor

$$\gamma = \frac{\pi^2}{3} k_B^2 D_F. \tag{5}$$

Furthermore, the term proportional to T^3 in (4) describes the contribution of acoustic phonons (see [5.3.2]). Via the Debye energy $k_B\theta_D = \hbar\omega_D$, A contains the atomic density (of the metal) and the mean sound velocity (of the acoustic phonons). Plotting $C_\Omega(T)/T$ versus T^2 (see fig. 5.1-1), a straight line with intercept γ and slope A is obtained. Thus γ and A, and consequently microscopic quantities D_F and θ_D or ω_D, can be determined from low-temperature measurements of $C_\Omega(T)$; the comparison of values determined theoretically and experimentally showing a discrepancy, which can be explained by the fact that the influence of the lattice oscillations on the electron states has been neglected in this consideration.

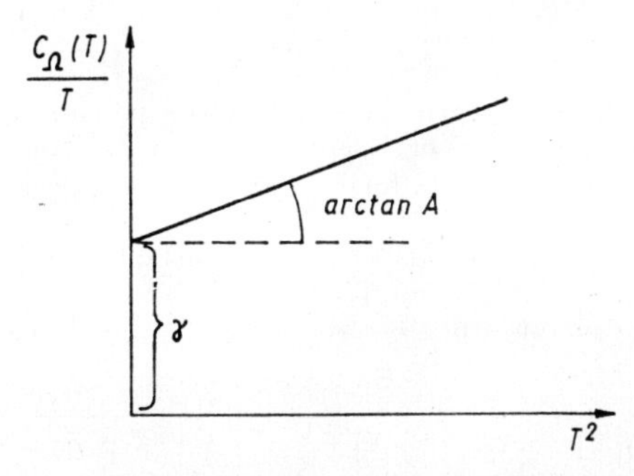

Fig. 5.1-1. Heat capacity of non-magnetic metals at low temperatures

5.1.2 Renormalization of Electron Energies

In the preceding sections, only the interaction of an electron with the static crystal potential and with the other electrons was taken into account. The corrections to the adiabatic approximation discussed in [1.3] lead to the electron-phonon interaction. According to [1.3], this interaction particularly modifies the behaviour of the electrons in the vicinity of the Fermi level, $|\varepsilon_{\boldsymbol{k},\nu} - \varepsilon_F| \sim \hbar\omega_D$. It can even lead to an effective attraction, which is the basis of the phonon mechanism of superconductivity. In an obvious physical picture, an electron (via Coulomb forces) polarizes the lattice in its surroundings. This distortion can in its turn act on other electrons. In a dynamic picture, the strength of the interaction is determined by the phonon energies of order $\hbar\omega_D$, which can be exchanged during this process.

The dynamic properties of an electron are also changed thereby, since the electron is compelled to drag its lattice polarization (or, in another representation, the cloud of virtual phonons it produces) with itself. This can best be described by a $\boldsymbol{k}$-dependent renormalization of the dispersion relation

$$\varepsilon_{\boldsymbol{k},\nu} \to \varepsilon_{\boldsymbol{k},\nu}/(1 + \lambda_{\boldsymbol{k},\nu}) . \tag{6}$$

Analogously, the DOS at the Fermi level proves to be

$$D_F \to D_F(1 + \lambda) \quad \text{with} \quad \lambda = \sum_{\boldsymbol{k},\nu} \lambda_{\boldsymbol{k},\nu}\delta(\varepsilon_{\boldsymbol{k},\nu} - \varepsilon_F)/\sum_{\boldsymbol{k},\nu} \delta(\varepsilon_{\boldsymbol{k},\nu} - \varepsilon_F) \tag{6'}$$

as the averaged electron-phonon coupling parameter. Its calculation is explained in [5.4]. It can be determined by means of tunnel spectroscopy or from the temperature T_c of the transition into the superconducting state. The orders of magnitude of D_F and λ for some metals are found in table 5.1-1.

The factor $(1 + \lambda)$ should also be taken into account in the electronic part of the heat capacity

$$\gamma = \frac{\pi^2}{3} k_B^2(1 + \lambda) D_F . \tag{5'}$$

Table 5.1-1. DOS D_F at the Fermi energy, electron-phonon coupling parameter λ, transition temperature of superconductivity T_c, and Debye energy $\hbar\omega_D$ for selected metals. The data for λ_{emp} are optimum values taken from tunneling and T_c-measurements after McMillan (68), Grimvall (81), Butler (77). λ_{theor} for simple metals see Grimvall (81); for transition metals: Papaconstantopoulos et al. (77) for 3*d*; Butler (77) for 4*d*; John et al. (82) for 5*d*. For simple metals, the corresponding r_s-values (see [1.1.2]) are given in brackets.

metal	D_F/ryd^{-1}	λ_{theor}	λ_{emp}	T_c/K	$\hbar\omega_D$/meV
Be (1.87)	0.80	0.26	0.24	0.03	89
Zn (2.29)	4.28	0.42	0.43	0.85	20
Pb (2.29)	7.10	1.47	1.55	7.19	9
α-Hg (2.65)	4.08	0.93	1.60	4.16	9
V	25.40	1.05	0.6	5.30	34
Nb	20.60	1.39	0.97	9.26	22
Ta	21.62	1.16	0.69	4.48	19

Particularly for the transition metals with their tendency to form long-range magnetic order, one has to remember that in the λ figuring in (5') a contribution caused by spin fluctuations is also contained, besides the electron-phonon part. This seems to apply particularly to the beginning of the transition period with its tendency to anti-ferromagnetic order (Sc, Y, Lu) and to the end, with the tendency to ferromagnetic order (Pd, Pt). The influence of the spin fluctuations on the electronic heat capacity can be suppressed by applying an external magnetic field.

The renormalization of the electron energies also manifests itself in other measurable quantities as cyclotron masses and amplitudes of dHvA oscillations. Quantities connected with the geometry of the Fermi surface only, such as the periods of dHvA oscillations, remain unchanged.

5.1.3 Transition Temperature of Superconductivity

The effective attraction of electrons near the Fermi energy brought about by electron-phonon interactions — provided that it exceeds the Coulomb repulsion of electrons there — leads to the formation of so-called Cooper pairs from electrons with opposite momenta and antiparallel spins. According to the theory of Bardeen, Cooper, and Schrieffer (BCS), all electrons in the vicinity of the Fermi level in a range with a width of the Debye energy $\hbar\omega_D$ condense into Cooper pairs. As distinct from $\varepsilon_F \approx 5 \cdots 10$ eV, the Debye energies amount to as little as some 10 meV: for specific examples for T_c and $\hbar\omega_D$, see table 5.1-1. To generate two free electrons, such a Cooper pair with a binding energy of 2Δ has to be broken up. Thus a forbidden zone of this width appears in the one-particle excitation spectrum. If these excitation energies are not available, a state of the system with total momentum different from zero and, thus, an electric current cannot be damped by the scattering of electrons, i.e., a superconducting state is present. With increasing temperature, the energy gap $\Delta(T)$ diminishes until it vanishes at the critical temperature T_c. Then the system takes on the normally conducting state. According to that BCS theory, for the energy gap $\Delta(0)$ at $T = 0$, it holds

$$2\Delta = 3.52 k_B T_c \tag{7}$$

and

$$k_B T_c = 0.86 \hbar\omega_D \, e^{-1/D_F V}. \tag{8}$$

Thus T_c increases with the magnitude $\hbar\omega_D$ of the exchanged energy. It also increases with the number of electrons participating in the pair formation in the vicinity of the Fermi level ($\sim D_F$), as well as the strength of the effectively attracting interaction V, which has been assumed in (8) to be given within a model and to be small.

The improved theory of Eliashberg (60) starts from the system of coupled equations for the Green functions of electrons and phonons. Its approximate numerical evaluation according to McMillan (68) yields the expression

$$k_B T_c = \frac{\hbar\langle\omega\rangle}{1.20} \exp\left[-\frac{1.04(1+\lambda)}{\lambda - \mu^*(1+0.62\lambda)}\right] \tag{9}$$

which is also valid for stronger interactions ($\lambda \approx 1$). Here $\langle\omega\rangle$ is a mean phonon frequency. The Coulomb pseudopotential μ^* globally takes into account the repulsive Coulomb interaction of the conduction electrons (for transition metals it is assumed to be $\mu^* \approx 0.13$, for simple metals somewhat smaller values are found). (8) is contained in (9) as a special case for small values of λ and $(\lambda - \mu^*)$.

Of the pure metals, Pb, α-Hg, Nb, and V (see table 5.1-1) exhibit maximum values of λ, whilst for most other metals λ is about 0.5. For the transition metals and their alloys, T_c shows a strong dependence on the number Z of valence electrons per atom (Matthias rule). Maxima of T_c are found for $Z = 4.5$ and 6.75. This tendency could be confirmed by calculating the dependence of λ along the 3*d*- and 4*d*-periods (Papaconstantopoulos et al. 77, Butler 77) as well as the 5*d*-period (John et al. 82) of the transition metals. Interpolation in the spirit of a rigid-band model explains the values for the alloys. Altogether these data show that high values of D_F do not necessarily correspond to high values of λ as shown by the examples of Pd and Pt, which show only small λ and thus no superconductivity in spite of a high D_F. At most, it holds vice versa that small values of D_F are coupled with small values of λ. To this end, we consider the elements Ta ($Z = 5$) and W ($Z = 6$). Both 5*d*-metals have the same lattice structure (*fcc*). The band structures and DOS are very much alike besides the fact that ε_F lies near a maximum of $D(\varepsilon)$ for Ta, while lying in a minimum for W ($D_F^W = 5$ per atom and *ryd*, $D_F^{Ta} = 22$ per atom and *ryd*). The corresponding values of λ and T_c are $\lambda^W = 0.3$, $\lambda^{Ta} = 0.7$, $T_c^W = 0.012$ K, $T_c^{Ta} = 4.48$ K. Analogous statements hold for the corresponding 4*d*-metals Nb ($Z = 5$, $\lambda = 0.97$, $T_c = 9.26$ K) and Mo ($Z = 6$, $\lambda = 0.44$, $T_c = 0.92$ K).

An important test for the theory is the pressure dependence of T_c. While for simple metals the influence of the pressure dependence of the phonon spectrum dominates and leads to $\partial T_c/\partial p < 0$, the positive values of $\partial T_c/\partial p$ observed at the beginning of the transition metal periods can be explained with the help of the theory treated in [5.4]. The physical explanation is based on the relatively large ionic radii of these elements.

Ferromagnetism as a cooperative phenomenon with different band structures for different spin directions competes with the superconductivity and its balance of electron spins in the Cooper pairs (see [3.2]). The possibility of their coexistence has been in dispute for a long time (Vonsovskij et al. 77). Therefore, recently there has been interest in ternary rare-earth compounds[1])

[1]) Examples are $(RE)M_4B_4$ with for example M = Rh, Ru, Ir, Os and RE = Er, Y; $(RE)Mo_6S_8$ with for example RE = Ho; and $(RE)Rh_xSn_y$, $Er_3Rh_4Sn_{13}$; $(RE)_2Fe_3Si_5$, $(RE)_5Co_4Si_{10}$, $(RE)Cu_2Si_2$. In these compounds the 4*f*-electrons of the RE ions carry the magnetic moments (localized at the corresponding lattice sites), whereas the 3*d*-, 4*d*-, and 5*d*-electrons of the transition metal ions are mainly responsible for the super-

as well as in the binary compound Y_9Co_7[1]) as "magnetic superconductors".[2])[3]) For UPt_3, one observes the coexistence of spin fluctuations and superconductivity. For Pd, also a sensitive interaction between both these influences is to be observed. On the one hand, as a consequence of a very narrow 4d-band the magnetic susceptibility is very high, which leads to strong spin fluctuations. On the other hand, the calculated λ is not as small as to exclude superconductivity at all. While Pd tends to ferromagnetic order, there is a tendency to antiferromagnetic order with Y. The competition between ferromagnetism and superconductivity is also found with the lanthanides.

5.1.4 Superconductors with High Critical Temperatures

After a long period of stagnation (Nb_3Ge — the compound with the highest T_c of 23.2 K — was already known in 1973!), Bednorz and Mueller (86) discovered a new class of superconductors with considerably higher critical temperatures T_c: The importance of this break-through was recognized by the awarding of a Nobel prize in physics only one year after the first publication! Currently, three groups of high-temperature superconductors are known (fig. 5.1-2):

(1) $(La, M)_2CuO_4$ with M = Ca, Sr, Ba (fig. b))
(K_2NiO_4 structure type), highest T_c = 35 K[4]) for $La_{1.85}Sr_{0.15}CuO_{4-y}$,
(2) $MBa_2Cu_3O_{7-y}$ with M = Sc, Y, La, Nd, Sm, Gd, Tb, Dy, Ho, Er, Tu, Yb and binary combinations with Lu; $y = 0 \cdots 0.5$ (denoted as 123 structure, fig. c)), $T_c \approx 95$ K (Wu et al. 87),
(3) $Bi_2Sr_2Ca_nCu_{n+1}O_{n+6+y}$; T_c = 110 K[4]) ($n = 2$), 85 K ($n = 1$), <20 K ($n = 0$), ($Tl_2Ca_2Ba_nCu_{n+1}O_{n+6+y}$; T_c = 127 K[4]) ($n = 2$), 105 K ($n = 1$), 80 K ($n = 0$), in these alloys perovskite and BiO layers alternate (see fig. d)).

They are oxidic systems with a perovskite-like structure which show metallic conductivity and a non-zero partial DOS of oxygen at the Fermi energy.

A common structure element of all three types is a CuO_2 layer (fig. 5.1-3) with a partially filled band which is formed by antibonding bonds between Cu 3d-states of symmetry ($x^2 - y^2$) and O 2p-states in the x- or y-direction. The role of the other layers is still unclear. Probably, they are important (i) as a polarizable medium which determines the lattice constant and separates the

conductivity. There are peculiarities of the superconductivity in $CeCu_2Si_2$ and UBe_{13} (heavy fermion systems). For further details see Eckern et al. (84).

[1]) Here magnetic order arises from itinerant electrons.

[2]) With the organic superconductors $(TMTSF)_2X$, X = ClO_4, PF_6, AsF_6, antiferromagnetic order appears above the transition temperature.

[3]) For the theory of magnetic superconductors see Tachiki (83).

[4]) The data on T_c differ somewhat because the transition curves of the electrical resistivity as a function of temperature have a width of a few K depending on the homogeneity of the samples.

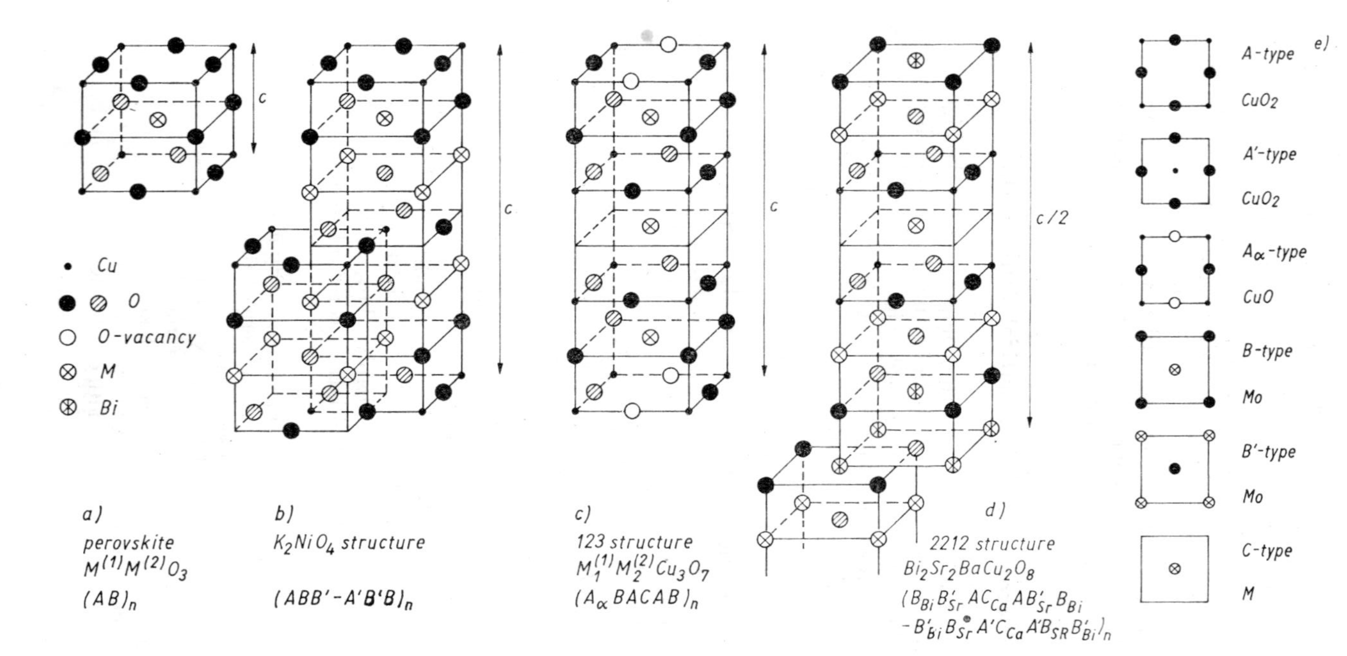

Fig. 5.1-2. Lattice structure of perovskite-like high-temperature superconductors. They result by stacking the different layers A, A′, A_α, B, B′, or C of fig. e). For simplicity the distances between the layers are assumed to be equal (c lattice constant in stacking direction)
a) perovskite structure b) K_2NiO_4 structure type, the M sites are occupied by La, Sr, Ba, or Ca atoms, after 3 layers the stack is shifted by (1/2, 1/2, 0) in the direction of the layer diagonal c) 123 structure, the layer A_α differs from the layer A by the missing oxygen atoms; therefore, it consists of Cu—O chains in y-direction d) 2212 structure, the CuO_2 layers are separated by Ca layers alternating with a group of SrO—BiO—BiO—SrO layers e) types of elementary layer A and A′ or B and B′, respectively, differ by a shift of half the diagonal (1/2, 1/2, 0) only. Type A_α results from A by the marked oxygen vacancies. C layers consist only of metal atoms without oxygen

CuO_2 layers, and (ii) for doping. T_c and other physical properties depend very strongly on oxygen content (denoted by the deviation y from the stoichiometry in the structure formulae). In particular, defects (vacancies) in the CuO chains in layers of type A_α determine the filling of the Cu $3d$—O $2p$ band. The layer structure is responsible for strong anisotropies in electrical conductivity and in the parameters of superconductivity.

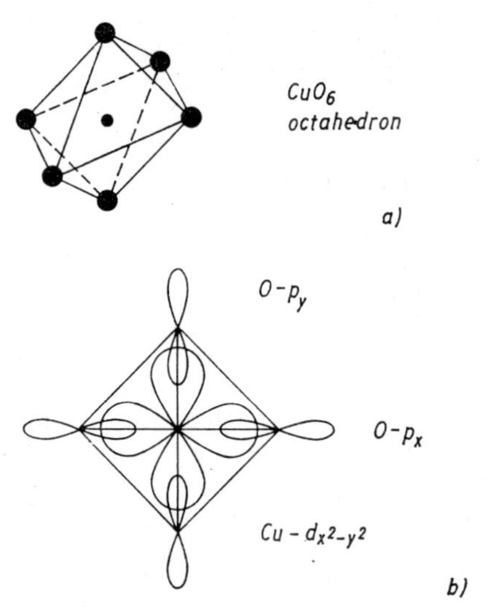

Fig. 5.1-3. a) CuO_6 octahedron as an element of short range order in perovskite-like structures. In the 123 and 2212 structure, one or two oxygen atoms are missing but the Cu atom is always surrounded by 4 atoms like in fig. b).
b) Bonding between $3d_{x^2-y^2}$ and $2p_x$, $2p_y$ states of the neighbouring Cu and O atoms, respectively

The mechanism of superconductivity is also still unclear. It has been shown experimentally that electron pairs are responsible for superconductivity. In comparison to known metallic systems, the non-vanishing DOS for oxygen sites at the Fermi energy is an important new feature. It implies that the oscillations of the (light!) oxygen atoms contribute to the electron-phonon coupling parameter λ in (5.4-9). Because of the relation $\langle\omega\rangle \sim M^{-\beta}$, a dependence on the oxygen mass (isotope effect) with exponent $\beta = 1/2$ is expected, according to (5.4-9). The observed isotope effect is much weaker ($\beta = 0.03 \cdots 0.05$ in 123 structures, Leary et al. 87). Therefore, the conclusion may be drawn that other, non-phononic mechanisms also contribute to λ. Other new features of the high-T_c superconductors to be taken into account in discussing the electron-phonon mechanism are the anisotropy and anharmonicty of the lattice vibrations. Quantitative estimations of these effects show that T_c of the K_2NiO_4 structure may be understood under these assumptions, but for the explanation of the experimental results in the 123 structures, a strong additional electronic correlation in the narrow Cu $3d$—O $2p$ band (with a band width $\approx 3 \cdots 5$ eV, smaller than in the known superconducting systems) is generally assumed. At present, a strong controversy on the actual form of these contributions exists.

Apart from a basic point of view, the new high-T_c superconductors are of great practical importance because they can be operated with liquid nitrogen as the cooling medium (cooling temperature $T_N = 77$ K), which reduces costs considerably. Usually it is assumed that for technical applications the critical temperature T_c should exceed the cooling temperature by factor 1.5 at least (i.e. $T_c \gtrsim 115$ K).

To this aim, the following problems must be solved:

- processing of flexible wires or tapes from brittle oxidic materials,
- increasing the density of the critical current of the material produced by ceramic technology (presently 10^3 A/cm^2) above values of 10^5 A/cm^2.

By means of molecular beam epitaxy, very thin crystalline layers with critical current densities of $1.5 \cdot 10^7$ A/cm^2 have been obtained. The observed critical magnetic fields H_c exceed considerably those of other known superconductors.

Several publications have reported superconductors with critical temperatures at or above room temperature. However, the results are not accepted by the physical community because either the superconducting properties are stable only for a short time, or else the experiments are not at all reproducible.

Enormous further development of this field is to be expected.

5.1.5 Magnetic Susceptibility

Another quantity determined by D_F is the magnetic susceptibility. The contributions to this quantity are: diamagnetism of closed-shell ion cores, Pauli spin paramagnetism of valence electrons, and Landau orbital diamagnetism of valence electrons.

Let us consider first spin paramagnetism. Its calculation by Pauli in 1927 proved to be the first application of quantum statistics to solid state physics. It results from the magnetic spin momenta of the electrons being adjusted themselves in an external magnetic field $\boldsymbol{B}$. The two possibilities for the spin are connected with energy shifts

$$\Delta\varepsilon = \pm\mu_s B \quad \text{for} \quad \begin{cases} \text{spin antiparallel to } \boldsymbol{B} \\ \text{spin parallel to } \boldsymbol{B} \end{cases}$$

related to the zero-field case (note that $\mu_s < 0$ because of $e < 0$: spin and magnetic momentum are opposite to each other). Therefore, in the case of free electrons, the band with $\varepsilon_{\boldsymbol{k}}^0 = (\hbar\boldsymbol{k})^2/2m$ and $D^0(\varepsilon) = D_F^0 \sqrt{\varepsilon/\varepsilon_F^0}$, $D_F^0 = 3/2\varepsilon_F^0$ is replaced by two bands shifted about $\pm\mu_s B$ with one half of the original DOS each and different spin directions (see fig. 5.1-4). Both bands are filled up to the common Fermi level at $T = 0$. The surplus electrons with momenta parallel to the field therefore equals the doubly hatched area

$$N^0_\pm = \int_{\pm\mu_s B}^{\varepsilon_F^0} d\varepsilon \, \frac{1}{2} \, D^0(\varepsilon \mp \mu_s B) \curvearrowright$$

$$N^0_+ - N^0_- = \int_{\varepsilon_F^0+\mu_s B}^{\varepsilon_F^0-\mu_s B} d\varepsilon \, \frac{1}{2} \, D^0(\varepsilon) \approx |\mu_s| \, B D^0(\varepsilon_F^0) . \tag{10}$$

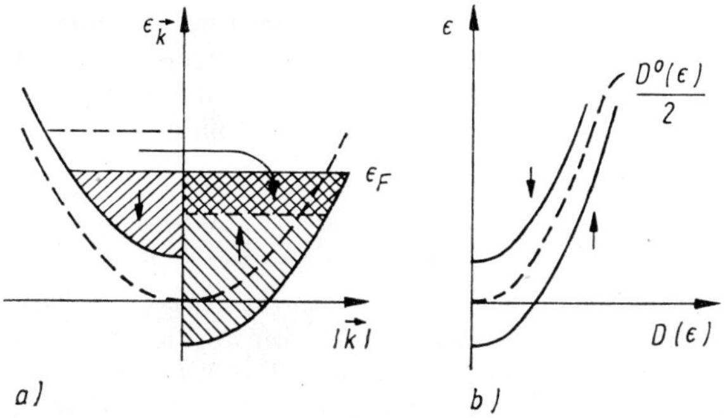

Fig. 5.1-4. Free electrons in a magnetic field, a) energy states, b) DOS

For the magnetization $M^0 = (N^0_+ - N^0_-) \, |\mu_s|/\Omega$, it therefore follows that $M^0 = \chi^0_{\mathrm{para}} \, B$ with

$$\chi^0_{\mathrm{para}} = \mu_s^2 \, \frac{D^0(\varepsilon_F^0)}{\Omega} \tag{11}$$

as the spin susceptibility. Thus it is (as is the electronic heat capacity) proportional to the DOS at the Fermi level. An analogous relation holds for Bloch electrons $\chi_{\mathrm{para}} = \mu_s^2 D(\varepsilon_F)/\Omega$. As a consequence of Fermi-Dirac statistics, χ_{para} is nearly independent of temperature. Incidentally, χ_{para} can also be determined directly from nuclear resonance measurements (Knight shift).

Finally the orbital diamagnetism explained by Landau in 1930 yields $\chi^0_{\mathrm{dia}} = -(1/3) \, \chi^0_{\mathrm{para}}$ (the negative sign results from the fact that the switching-on of the magnetic field is connected with an energy increase for the electrons), free electrons thus behaving paramagnetically altogether and contributing $(2/3) \, \chi^0_{\mathrm{para}}$ to the magnetic susceptibility. For Bloch electrons the theory of orbital diamagnetism is more complicated. An additional effect is the above-mentioned influence of the ion cores. This is why measurements of susceptibility are less appropriate for the determination of D_F. Furthermore, one has to take into account the renormalization of the susceptibility by the electron-electron interaction (see Moruzzi et al. 78, Ashcroft and Mermin 76)[1]).

5.1.6. Transport Quantities

Many theoretical investigations of transport coefficients, such as electrical and thermal conductivity, and also thermopower, magnetoresistivity (transverse, longitudinal), Hall coefficient, etc. start from the quasiclassical concept of the Boltzmann equation, describing the system of valence electrons by a one-partial distribution function $f(\boldsymbol{r}, \boldsymbol{p}, t)$ or $f_{\boldsymbol{k}}(\boldsymbol{r}, t)$, which follows from the unperturbed Fermi-Dirac distribution (1.1-7) under the influence of small perturbations, which may depend on time and, weakly, on position. The simultaneous information on position $\boldsymbol{r}$ and momentum $\boldsymbol{p}$ is not in contradiction with quantum mechanics since we can assume that the wave packets — as discussed in [4.1] — are much smaller than the regions in which the external perturbations and thus $f_{\boldsymbol{k}}(\boldsymbol{r}, t)$ change to any extent. As mean particle density in phase space, $f_{\boldsymbol{k}}(\boldsymbol{r}, t)$ is normalized to the total number of valence electrons.

The Boltzmann equation reads

$$\left(\frac{\partial}{\partial t} + \boldsymbol{v}_{\boldsymbol{k}} \frac{\partial}{\partial \boldsymbol{r}} + \boldsymbol{F}_{\boldsymbol{k}} \frac{\partial}{\partial \hbar \boldsymbol{k}}\right) f_{\boldsymbol{k}}(\boldsymbol{r}, t) = \left(\frac{\partial f_{\boldsymbol{k}}}{\partial t}\right)_{\text{coll}} \tag{12}$$

with $\boldsymbol{v}_{\boldsymbol{k}} = \partial \varepsilon_{\boldsymbol{k}}/\partial \hbar \boldsymbol{k}$ as the electron velocity according to (4.1-3) and $\boldsymbol{F}_{\boldsymbol{k}} = e(\boldsymbol{E} + \boldsymbol{v}_{\boldsymbol{k}} \times \boldsymbol{B})$ as the external force on an electron. For the sake of simplicity, only one energy band is considered and, therefore, the band index is dropped, hence $k = (\boldsymbol{k}, \nu) \to \boldsymbol{k}$. Equation (12) proves to be a balance equation in phase space, its left hand side (multiplied by $\mathrm{d}^3r\, \mathrm{d}^3p\, \mathrm{d}t$) means the total change of the number of electrons in the phase space volume $\mathrm{d}^3r\, \mathrm{d}^3p$ during $\mathrm{d}t$, if this volume moves with the phase space points according to $\boldsymbol{v}_{\boldsymbol{k}}$ and $\boldsymbol{F}_{\boldsymbol{k}}$. For a volume $\mathrm{d}^3r\,\mathrm{d}^3p$ fixed in phase space, $\mathrm{d}^3r\,\mathrm{d}^3p\,\mathrm{d}t\,(\boldsymbol{v}_{\boldsymbol{k}}\partial/\partial\boldsymbol{r} + \boldsymbol{F}_{\boldsymbol{k}}\partial/\partial\boldsymbol{p}) f_{\boldsymbol{k}}$ is the number of particles which leave this volume through its surface owing to their unperturbed motion in phase space. The change of the particle number due to interaction ('collisions'), e.g. with defects, lattice vibrations, or other electrons, which is important for the evolution towards the equilibrium state, is described by the r.h.s., the so-called 'collision term', which contains the transition probabilities (i.e. the probabilities per time) $P_{\boldsymbol{k},\boldsymbol{k}'}$, for scattering from $\boldsymbol{k}$ to $\boldsymbol{k}'$ (or vice versa because of microscopic reversibility), and, in its turn, describes a balance:

$$\left(\frac{\partial f_{\boldsymbol{k}}}{\partial t}\right)_{\text{coll}} = \sum_{\boldsymbol{k}'} P_{\boldsymbol{k},\boldsymbol{k}'}(f_{\boldsymbol{k}'} - f_{\boldsymbol{k}}), \tag{13}$$

[1]) On the contrary there is no renormalization of χ^0_{para} due to the electron-phonon interaction. The one-particle states are renormalized according to (6) within a range of $\hbar\omega_{\text{D}}$ on both sides of the Fermi energy. This shell 'rides' on top of the Fermi sea and can therefore not be 'filled up' with electrons that turn their spins under anapplied magnetic field (see Grimvall 81, p. 140).

Thus the theoretical investigation of λ is mainly concentrated on calculation of η. Since the integrands in (8) vanish outside the MT sphere, η can be represented by scattering phase shifts $\delta_l = \eta_l(\varepsilon_F)$ and the Bloch functions can be expressed within a lattice harmonic expansion (see also (2.5-11))

$$\psi_k(\boldsymbol{r}) = \sum_{l,\alpha,\mu} i^l C^\alpha_{l\mu}(k)\, R_l(r, \varkappa)\, K^\alpha_{l\mu}(\boldsymbol{r}). \tag{11}$$

The coefficients $K^\alpha_{l\mu}(\boldsymbol{r})$ are real linear combinations of spherical harmonics, which transform themselves like the irreducible representations of the point group of the lattice site considered. α denotes the irreducible representation (e.g. Γ_1, Γ_2, Γ_{15}, Γ_{12} for cubic symmetry) and μ is the number of functions belonging to the irreducible representation α. The coefficients $C^\alpha_{l\mu}(k)$ are required in the following combination:

$$\operatorname{Im} A^\alpha_{ll'}(\varepsilon) = \frac{\hbar^2}{2m}\,\frac{\pi}{\varkappa} \sum_k i^{l-l'} C^\alpha_{l\mu}(k)\, C^\alpha_{l'\mu}(k)\, \delta(\varepsilon - \varepsilon_k) \tag{12}$$

(they may be chosen real for lattice structures with inversion symmetry).[1]) The diagonal elements $l' = l$ determine the partial DOS within the MT sphere with respect to the angular momentum l and the irreducible representation α (dimensionality g^α):

$$D^\alpha_l(\varepsilon) = [\operatorname{Im} A^\alpha_{ll}(\varepsilon)]\, g^\alpha D^0_l(\varepsilon), \quad D^0_l(\varepsilon) = \frac{2m}{\hbar^2}\,\frac{\varkappa}{\pi} \int_0^{R_{\mathrm{MT}}} \mathrm{d}r r^2 R^2_l(r, \varkappa) \tag{13}$$

(note that $R_l(r, \varkappa)$ depends on ε via $\varkappa = \sqrt{2m\varepsilon/\hbar^2}$). $D^0_l(\varepsilon)$ is the partial DOS of an isolated MT potential, and therefore $\operatorname{Im} A^\alpha_{ll}$ is a direct measure of the influence of the lattice structure and the multiple scattering effects on the partial DOS. For cubic systems (*bcc*, *fcc*), η separates into spherical (η_0) and non-spherical (η_1) contributions, $\eta = \eta_0 + \eta_1$ (to avoid confusion between the Hopfield-parameter and its contributions with the scattering phase shifts, the latter are denoted by δ_l). The first term is given by

$$\eta_0 = \frac{2}{D_F}\,\frac{\varkappa_F^2}{\pi^2} \sum_l 2(l+1) \sin^2(\delta_l - \delta_{l+1})\, \bar{A}_l \bar{A}_{l+1} \tag{14}$$

where $\bar{A}_l = \sum_\alpha g^\alpha \operatorname{Im} A^\alpha_{ll}(2l+1)$. η_0 does not depend explicitly on the lattice structure and is the main contribution to η. The non-spherical correction η_1 depends explicitly on the lattice structure (John et al. 78 and references therein).

[1]) In the electron theory of point defects (see [5.5.])

$$\operatorname{Re} A^\alpha_{ll}(\varepsilon) = \frac{P}{\pi} \int \mathrm{d}\varepsilon' \frac{1}{\varepsilon' - \varepsilon} [\operatorname{Im} A^\alpha_{ll}(\varepsilon') - 1] \frac{D^0_l(\varepsilon')}{D^0_l(\varepsilon)} \tag{12'}$$

is also needed, where P means the Cauchy principal value.

For vanadium, for example, it was found that $\eta_0 = 6.07$ eV/Å^2, $\eta_1 = 0.17$ eV/Å^2, and hence $\eta = 6.24$ eV/Å^2, where η_0 is mainly determined by the term $l = 1$ of (14) contributing 4.0 eV/Å^2. The latter can be reduced by half taking into account screening (Pettifor 77). This improves the agreement between the theoretical (1.05) and experimental (0.60) λ value. Further results for λ are given in Table 5.1-1. For heavier elements with atomic numbers greater than 50, one has to take into account relativistic effects in band structure calculations. This corrects η_0 for example in the case of Ta, Ir, Pt by $+2\cdots6\%$ only (John et al. 82), but becomes essential for the actinides (Skriver and Mertig 85, and Skriver et al. 88).

$\alpha^2(\omega)\,F(\omega)$ was calculated by Butler et al. (79) and Pinski et al. (81) within the RMTA for Nb and Pd. These results have shown that $\alpha^2(\omega)$ is only to a very poor approximation independent of ω (see also Butler 81). λ and T_c were calculated by Glötzel et al. (79) for all 4d- and 5d-transition metals starting from $\alpha^2(\omega)\,F(\omega)$ by numerical solution of the Eliashberg equation (see also [5.1.3]).

Also within the RMTA, calculated phonon life times agree well with experimental data for Nb (Butler et al. 79, Butler 81). Although the RMTA describes some trends of the electron-phonon interaction of the transition metals correctly, the calculation of the electron-phonon matrix element from first principles still remains an interesting problem.

5.5 Electronic Properties of Localized Defects

Electronic properties (e.g. residual resistivity) of point defects (e.g. 3d-transition metal impurities in Al or Cu) can be calculated from first principles (see e.g. Mertig et al. 87, Mrosan and Lehmann 83).

5.5.1 Defects, Physical Properties, and Simple Model Conceptions

Perturbations of the ideal crystal give rise to a finite life time of the electron states (characterized by the quasimomentum $\hbar\boldsymbol{k}$) and thus cause physical effects such as residual resistivity. Such perturbations include, in addition to phonons, all kinds of defects, which appear in non-ideal crystals (vacancies including voids, self-interstitials, substitutional and interstitial impurities, dislocations, grain boundaries and interfaces). Point defects are the simplest case. For point defects, there are very detailed investigations (see e.g. Jena 81, Mertig et al. 87). From a theoretical point of view, localized point defects can be described in a relatively simple way. For simplification, it is assumed that the perturbation caused by a point defect is localized within the WS cell where the point defect is located. In the following, we only consider substitutional impurities (dilute alloys: in this limit the impurity induced physical properties vary linearly as a function of impurity concentration). We assume the point defect to be non-magnetic and we limit ourselves to the case $T = 0$.

As a basis for the theoretical description, we take the model of non-overlapping spherically symmetric MT potentials. The potential of the perturbing impurity is described by scattering phase shifts $\eta_l^i(\varepsilon)$, the potential of the host atoms by $\eta_l^h(\varepsilon)$. The method of multiple scattering theory is used (for the details see Mertig et al. 87). For a description of localized impurities on the basis of the tight-binding (TB) model, refer to Mrosan and Lehmann (83), section 11.4.

The following impurity induced physical properties are considered: change of DOS, change of electronic specific heat, change of charge density, Dingle temperature, residual resistivity, impurity thermopower, magnetoresistivity and Hall coefficient due to impurities.

5.5.2 Neglecting the Host Band Structure (free electron model)

Firstly we neglect the host band structure by using $\eta_l^h(\varepsilon) = 0$, and seek only the influence of a single potential perturbation of MT type on a gas of free non-interacting valence electrons (Friedel 52).

The integrated DOS is changed by[1])

$$\Delta N(\varepsilon) = \frac{2}{\pi} \sum_l (2l + 1) \, \eta_l^i(\varepsilon) . \tag{1}$$

ΔZ is the valence difference connected with the impurity induced potential perturbation. The charge neutrality of the system, i.e. the screening of the impurity induced potential perturbation by the host conduction electrons, is described by the Friedel sum rule $\Delta Z = \Delta N(\varepsilon_F)$. The derivative of (1) with respect to energy gives the change of the DOS $\Delta D(\varepsilon)$ which, according to (5.1-5), is directly related to the corresponding change in coefficient of the electronic specific heat.

The impurity induced potential perturbation further causes a change in particle density of the electron gas: in the asymptotic region the well known Friedel oscillations $\Delta n(r) \approx (A/r^3) \cos (2k_F r + \Phi)$ are obtained, where both amplitude A and phase Φ are determined by the scattering phase shifts at the Fermi energy $\eta_l^i(\varepsilon_F)$.

According to the discussion in [5.1.6], the residual resistivity can be obtained, if the scattering probability $P_{\boldsymbol{k},\boldsymbol{k}'}$ is known by solving the Boltzmann equation. The golden rule gives

$$P_{\boldsymbol{k},\boldsymbol{k}'} = N_i \frac{2\pi}{\hbar} |T_{\boldsymbol{k},\boldsymbol{k}'}|^2 \, \delta(\varepsilon_{\boldsymbol{k}} - \varepsilon_{\boldsymbol{k}'}) \tag{2}$$

[1]) The increase of the phase shift $\eta_i^i(\varepsilon)$ by π means the wave function has an additional node i.e. there is — compared with the unperturbed system — just one additional state at energy ε and angular momentum l. Of course the electron contributes to all angular momentum states. The sum over all angular momentum and spin contributions yields just (1).

with

$$T_{\boldsymbol{k},\boldsymbol{k}'} = -\frac{4\pi}{\Omega}\frac{\hbar^2}{2m}\frac{1}{k_\mathrm{F}}\sum_l (2l+1)\sin\eta_l^i(\varepsilon_\mathrm{F})\,\mathrm{e}^{i\eta_l^i(\varepsilon_\mathrm{F})}P_l(\cos\vartheta) \tag{3}$$

where N_i is the number of impurity atoms and $T_{\boldsymbol{k},\boldsymbol{k}'}$ is the T matrix for the impurity induced potential perturbation $V^i(r)$ of the free electron gas. It is remarkable that all integrals of type (2.5-12), necessary for obtaining (3), according to (2.5-14) can be transformed into expressions which only contain the phase shifts. Substituting (2) and (3) into (5.1-22), according to (5.1-20) one obtains for the residual resistivity

$$\varrho = \varrho(\varepsilon_\mathrm{F}) = \frac{N_i}{ZN_a}\frac{\hbar}{e}\frac{4\pi}{k_\mathrm{F}}\sum_l (l+1)\sin^2[\eta_l^i(\varepsilon_\mathrm{F}) - \eta_{l+1}^i(\varepsilon_\mathrm{F})] \tag{4}$$

where ZN_a is the number of valence electrons (Z = valence of a host atom; N_a = number of all atoms). The coupling of neighbouring angular momenta (l and $l+1$) is due to the transport factor $(1 - \cos\vartheta)$ in formula (5.1-22). From $\varrho(\varepsilon)$ also the thermopower is obtained according to

$$Q = Q(\varepsilon_\mathrm{F}) = \frac{\pi^2}{3}\frac{k_\mathrm{B}}{|e|}k_\mathrm{B}T\left.\frac{\mathrm{d}\ln\varrho(\varepsilon)}{\mathrm{d}\varepsilon}\right|_{\varepsilon_\mathrm{F}} \tag{5}$$

(Mott formula, see for example Ziman 76).

The dHvA oscillations are damped for non-zero temperatures or for non-zero concentration of point defects. The influence of point defects is described in terms of the so-called Dingle temperature T_D (this is the fictive temperature $T > 0$ of a corresponding ideal host for giving the same damping of the dHvA oscillations as a finite concentration of point defects in the same host at $T = 0$). One gets

$$k_\mathrm{B}T_\mathrm{D} = -N_i\frac{1}{\pi}\oint\frac{\mathrm{d}k}{v_{\boldsymbol{k},\perp}}\,\mathrm{Im}\,T_{\boldsymbol{k},\boldsymbol{k}}\bigg/\oint\frac{\mathrm{d}k}{v_{\boldsymbol{k},\perp}}. \tag{6}$$

The integrations in (6) have to be performed over the orbits of the electrons along the correspondingly considered external cross-sectional orbits of the Fermi surface. Here $v_{\boldsymbol{k},\perp}$ denotes the component of Fermi velocity in the plane of the orbit which is perpendicular to the external magnetic field. According to the optical theorem, $\mathrm{Im}\,T_{\boldsymbol{k},\boldsymbol{k}}$ is proportional to the scattering cross-section and in this way $\mathrm{Im}\,T_{\boldsymbol{k},\boldsymbol{k}}$ characterizes the scattering of the electron along the orbit. With (3) one obtains

$$k_\mathrm{B}T_\mathrm{D} = \frac{N_i}{ZN_a}\frac{4}{3\pi^2}\varepsilon_\mathrm{F}\sum_l (2l+1)\sin^2\eta_l^i(\varepsilon_\mathrm{F}). \tag{7}$$

All the physical properties considered are determined by the scattering phase shifts or their energy derivatives, both taken at the Fermi energy.

5.5.3 Consideration of the Host Band Structure

All impurity induced physical properties are changed, if according to $\eta_l^h(\varepsilon) \neq 0$ the host band structure is taken into account (neglected in the previous section). Instead of (1) one has

$$\Delta N(\varepsilon) = \frac{2}{\pi} \sum_{\alpha,\mu} \Phi_\alpha(\varepsilon), \tag{8}$$

$$\tan \Phi_\alpha(\varepsilon) = \frac{\operatorname{Im} A_{ll}^\alpha(\varepsilon)}{\cot \Delta\eta_l(\varepsilon) - \operatorname{Re} A_{ll}^\alpha(\varepsilon)} \tag{9}$$

where $\Phi_\alpha(\varepsilon)$ are the effective phase shifts, $\Delta\eta_l(\varepsilon) = \eta_l^i(\varepsilon) - \eta_l^h(\varepsilon)$ and Im, Re $A_{ll}^\alpha(\varepsilon)$ are defined in (5.4-12 and 12′), where also the meaning of α and μ is explained. Here only angular momenta $l \leq 2$ are considered which correspond to $\alpha = \Gamma_1$ $(l = 0)$, Γ_{15} $(l = 1)$, $\Gamma_{25'}$ $(l = 2)$ and Γ_{12} $(l = 2)$. Instead of (3) one gets

$$T_{\boldsymbol{k},\boldsymbol{k}'} = -\frac{\hbar^2}{2m} \frac{1}{\varkappa_\mathrm{F}} \sum_{\alpha,\mu} C_{l\mu}^\alpha(\boldsymbol{k})\, S_\alpha C_{l\mu}^\alpha(\boldsymbol{k}') \tag{10}$$

with $C_{l\mu}^\alpha(\boldsymbol{k})$ according to (5.4-11) and

$$S_\alpha = \frac{\sin \Phi_\alpha(\varepsilon_\mathrm{F})\, \mathrm{e}^{i\Phi_\alpha(\varepsilon_\mathrm{F})}}{\operatorname{Im} A_{ll}^\alpha(\varepsilon_\mathrm{F})} = \frac{1}{\dfrac{1}{\sin \Delta\eta_l(\varepsilon_\mathrm{F})\, \mathrm{e}^{i\Delta\eta_l(\varepsilon_\mathrm{F})}} - A_{ll}^\alpha(\varepsilon_\mathrm{F}) + i}\,. \tag{11}$$

Instead of (7) one obtains

$$k_\mathrm{B} T = N_i \frac{1}{\pi} \frac{\hbar^2}{2m} \frac{1}{\varkappa_\mathrm{F}} \sum_\alpha \frac{\sin^2 \Phi_\alpha(\varepsilon_\mathrm{F})}{\operatorname{Im} A_{ll}^\alpha(\varepsilon_\mathrm{F})} \oint \frac{\mathrm{d}k}{v_{\boldsymbol{k},\perp}} \times \sum_\mu |C_{l\mu}^\alpha(\boldsymbol{k})|^2 \Big/ \oint \frac{\mathrm{d}k}{v_{\boldsymbol{k},\perp}}\,. \tag{12}$$

For $\eta_l^h(\varepsilon) \to 0$ it follows $C_{l\mu}^\alpha(\boldsymbol{k}) \to 4\pi Y_{lm}(\boldsymbol{k})/\sqrt{\Omega}$. According to (5.4-12 and 12′), one gets $A_{ll'}^\alpha(\varepsilon) \to i\delta_{ll'}$ and (8) to (12) becomes the corresponding expressions of [5.5.2]. For (12) one must recall $\Omega k_\mathrm{F}^3/ZN_a = 1/3\pi^2$. Substituting (10) into (2) for the transition probability instead of (3), the following expression is obtained (Lehmann 73a):

$$P_{\boldsymbol{k},\boldsymbol{k}'} = \sum_m \alpha_m(\boldsymbol{k})\, \beta_m(\boldsymbol{k}')\, \delta(\varepsilon_{\boldsymbol{k}} - \varepsilon_{\boldsymbol{k}'})\,. \tag{13}$$

m is shorthand notation for l, α, μ, l', α', μ'. It is important that in (13) the dependence on $\boldsymbol{k}$ and $\boldsymbol{k}'$ has a factorized form. Since the sum over m comprises a finite number of terms only, (5.1-16) becomes an integral equation with a degenerate kernel and can be reduced to a linear algebraic set. With the ansatz

$$\Lambda_{\boldsymbol{k}} = \tau_{\boldsymbol{k}} \left[\boldsymbol{v}_{\boldsymbol{k}} + \sum_m \alpha_m(\boldsymbol{k})\, \boldsymbol{X}_m\right] \tag{14}$$

(5.1-16) becomes a linear algebraic set, and its solution gives $\boldsymbol{X}_m$, from which according to (14) $\Lambda_{\boldsymbol{k}}$ is obtained and finally according to (5.1-19 and 20) the residual resistivity is calculated. If the host band structure is taken into account only partially in terms of effective phase shifts $\Phi_\alpha(\varepsilon) = \Phi_l(\varepsilon)$ (spherical band approximation $A_{22}^{\Gamma 25'} = A_{22}^{\Gamma 12}$), an approximate expression in analogy to (4) is obtained,

$$\varrho \approx \varrho_0 \sum_l (l+1) \sin^2 [\Phi_l(\varepsilon_F) - \Phi_{l+1}(\varepsilon_F)]. \tag{15}$$

Because of the degenerate kernel (13), using Fermi surface harmonics, we also can calculate magnetoresistivity (longitudinal, transversal) and Hall coefficient as a function of the strength of the external magnetic field (Kohler diagram) (Mertig 82, Mertig and Mrosan 82, 83).

5.5.4 Results for the Residual Resistivity

For the calculations themselves, the impurity scattering phase shifts $\eta_l^i(\varepsilon)$ are needed, which follow from the impurity potential $V^i(r)$. The latter can be constructed in a self-consistent way on the basis of multiple scattering theory (Zeller and Dederichs 79). A simpler procedure is a model description including an open parameter, which is then fixed in an appropriate way.

A suitable procedure is to construct the potential at the impurity lattice site according to the Mattheiss prescription for which atomic charge densities of neutral atoms (impurity atom and neighbouring host atoms) are superimposed. This potential is then shifted by a constant value Δ (this is the above-mentioned parameter) in such a way that the Friedel sum rule is fulfilled (Mrosan and Lehmann 76). Here, charge transfer and screening effects are approximately described.

Using this procedure, the scattering phase shifts $\eta_l^i(\varepsilon)$ for $l = 0, 1, 2$ for $3d$-transition metal impurities (Sc, ..., Cu) were calculated (Mrosan and Lehmann 78). In each case, the d-phase shift $\eta_2^i(\varepsilon)$ is of special interest. According to (2.3-8), the d-phase shift is characterized by a resonance (the value $\pi/2$ is crossed at ε_d within an energy range of width Γ), which corresponds to a virtual bound state at ε_d with a lifetime $\hbar/\Gamma$. Table 5.5-1 shows the moving down and narrowing, which occur simultanously if the virtual bound state runs through the $3d$- series (both results are in agreement with photoelectric spectroscopy data). If ε_d is very close to the host Fermi energy (zero of $\varepsilon_d - \varepsilon_F$ nearly in the middle of the $3d$-series), the host conduction electrons are scattered very strongly (resonance scattering). In contrast, the scattering is the weaker, the greater the separation of ε_d and ε_F. This is the basis for understanding the trend of the residual resistivity with a maximum almost in the middle of the $3d$-series. Figure 5.5-1 shows the analogous situation for $3d$-transition metal impurities in Cu. The theoretical values were calculated by solving (14) as well as by using the approximation (15). There are two reasons for the fact that

Table 5.5-1. Position ($\varepsilon_d - \varepsilon_F$) and width Γ of the resonance in the phase shifts $\eta_2^i(\varepsilon)$ of $3d$-transition metal impurities in Al

point defect	$(\varepsilon_d - \varepsilon_F)$/eV	2Γ/eV
V	0.17	3.26
Cr	−0.25	2.56
Mn	−0.60	2.05
Fe	−1.03	1.63
Co	−1.84	1.25
Ni	−3.34	0.80
Cu	−6.57	0.30

the experimental residual resistivities in most cases are larger than the theoretical values, even with the host band structure taken into account:

(1) the (electronic) perturbation of the crystal potential at the sites of neighbouring host atoms is neglected;
(2) atomic lattice relaxations, which are connected with the incorporation of the defect, are neglected.

Both effects, of course, contribute to the residual resistivity.

The main contribution to ϱ comes from $\eta_2^i(\varepsilon_F)$. If by estimate the contributions from angular momenta $l = 0$ and $l = 1$ are neglected, the Friedel sum rule simply becomes $\Delta Z = (2/\pi)\, 5\eta_2^i(\varepsilon_F)$, or $\eta_2^i(\varepsilon_F) = (\Delta Z/10)\, \pi$, from which follows $\varrho \sim \sin^2 (\Delta Z/10)\, \pi$ according to (4). In this way, ϱ has a maximum for $\Delta Z = 5$ (almost in the middle of the $3d$-series) and vanishes for $\Delta Z = 0$ and 10.

The other defect induced physical properties (electronic specific heat, thermoelectric power, Dingle temperature, magnetoresistivity, Hall coefficient) show an analogous trend along the $3d$-series (Mrosan and Lehmann 76, Mertig 82, Mertig and Mrosan 82, 83). The reason is again the resonance scattering of the host conduction electrons at the virtual bound state due to the $3d$-electrons of the impurity. The concept of resonance scattering is also the basis for understanding the residual resistivity and of other physical properties of $4d$- and $5d$-transition metal impurities in Al and other hosts (Cu, Ag, Au). — A recent review on the electronic structure of impurities in transition metals is given by Zeller (87a). — An example for a recent paper dealing with light impurities (He, Li, ..., Ne) in Al and taking into account lattice relaxations is Perrot and Rasolt (88).

Extending the theory presented to the spin-polarized case, it is possible to calculate magnetic moments of impurities. The calculated magnetic moments for different impurities ($3d$, $4d$) in different hosts (Cu, Ag, Ni, Mo) agree very well with experimental data (Podloucky et al. 80, Braspenning et al. 84, Blügel et al. 87, Stefanou et al. 87, Ellialtioglu et al. 87, Zeller 87b). For vacancies and magnetic impurities in transition metals see Zeller (87c).

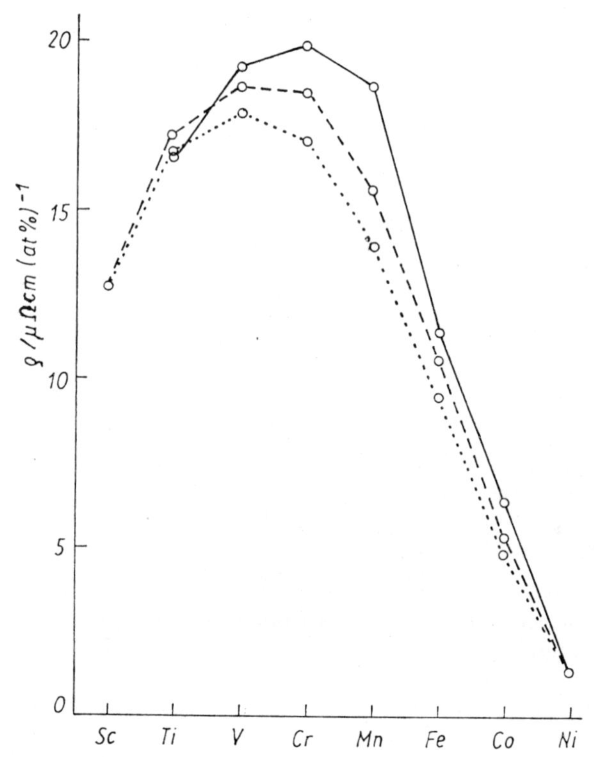

Fig. 5.5-1. Residual resistivity of $3d$-transition metal impurities in Cu (··· calculated with (14), --- calculated with (15), —— experimental values) after Mertig et al. (82)

5.6 Some Other Properties

The list of electronic properties treated in Chapters [4.1]—[4.3] and [5.1]—[5.5] is by no means complete. Some additional ones are mentioned in the following.

The gyromagnetic factor requires a careful relativistic treatment (Schober et al. 86). In nuclear magnetic resonance measurements and hyperfine interactions (Winter 71), i.e. in the nuclear spin-lattice relaxation time (Asada et al. 81) and in the Knight-shift (Ebert et al. 86), as well as in the coefficient of inner conversion (Grechukhin and Soldatov 86), and in the Mößbauer effect (Shenoy and Wagner 78) the nuclei appear as a probe of the electronic structure. With the transient thermoreflectance spectroscopy further non-equilibrium properties of metals are studied.

Interfaces, cohesion, adhesion, hydrogen and helium embrittlement, (brittle and ductile) fracture are hudge fields.

In the following surfaces and other low dimensional systems are studied.

6 Surfaces, Thin Layers, Quasi-One- and Quasi-Two-Dimensional Conductors

Systems with lower symmetry than three-dimensional solids are of increased interest. The reason is that the increasing understanding of infinitely extended systems gives better preconditions for the investigation of perturbations originating from their boundaries in the form of surfaces, but also of interfaces. Detailed theoretical investigations started at the end of the sixties. They are computationally expensive and need strong connection with experiment. The development of recent methods for the characterization of surfaces, such as photoelectron and Auger spectroscopy, ESCA, the diffraction of low-energy or reflection of high-energy electrons (LEED and RHEED, respectively), tunneling spectroscopy, scanning tunneling microscopy, or field emission of electrons and ions, and the availability of synchrotron radiation, has been a strong stimulus on the theoretical approaches.

Generally, bulk properties are already observed at a distance of 5 to 10 atomic layers from the surface. Therefore, considerable effects are to be expected only if the thickness of thin layers or the diameter of small particles are of this order. Moreover, as many technologically important processes are influenced considerably by the existence, formation, and change of surfaces and interfaces, the systematic investigation of related effects on a microscopic basis will develop rapidly both with respect to basic and also to applied research. Some examples for this trend are: heterogeneous catalysis, intercrystalline fracture, cutting metal working, properties of surface coatings, effects of VLSI techniques in microelectronics, epitaxy, preparation and stabilization of heterogeneous structures, and others. All these phenomena need a better understanding of properties of surfaces and interfaces. Presently, there is still a broad gap between the needs of materials science on the one hand and experimental and theoretical basic research on the other hand.

However, in the near future, as regards preparation and characterization of defined interfaces and surfaces self-consistent treatment of the electron-electron interaction within them understanding of physical and chemical processes like absorption and the basic processes on and in surfaces and interfaces, considerable progress is expected with respect to improvement of surface properties.

Moreover concerning systems with one- or two-dimensional structures, new physical effects are expected as a consequence of the reduced dimensionality.

Small particles, i.e. clusters of a few up to 1000 atoms, are characterized by curved and closed surfaces. Therefore, their properties are dominated by the surfaces the more the smaller they are. Small particles are studied experimentally and theoretically, this branch of physics and chemistry (and technology) is rapidly developing (Rao and Jena 87). — In a sense the opposite are vacancy clusters or voids (with negatively curved surfaces) within an otherwise perfect crystal. They are also studied intensively. For the atomic relaxation around such clusters in molybdenum see e.g. Khanna et al. 88.

6.1 Peculiarities of Surfaces and Thin Films

6.1.1 Characterization of Structure

The characterization of real surfaces is a very complicated problem since even under nearly ideal conditions during the preparation process, there are changes in comparison with the bulk both in the physical structure (distribution of atoms) and in the chemical structure, such as addition of atoms, apart from discussing the macroscopic deviations from a given mathematically rigorous surface shape, such as in the form of roughness. Therefore, at first we consider in fig. 6.1-1 a few of the possible microscopic deviations from plane surfaces.

(a) The case most favourable for a theoretical treatment appears if the pure surface exhibits the same periodicity as the bulk material. Practically, changes in the distance of the uppermost crystal layers always appear in comparison with the distances within the crystal. Also a parallel shift of all atoms within the surface layer may be observed.

(b) Within the surface layer (and within the neighbouring layers), *superstructures* are possible which originate — as indicated schematically in fig. b) — by different shifts of the atoms within the layer, if the new configuration is thermodynamically stable. Such a superstructure with an elementary cell eight times larger than in the bulk material has been observed in the (111) surfaces of germanium single crystals.

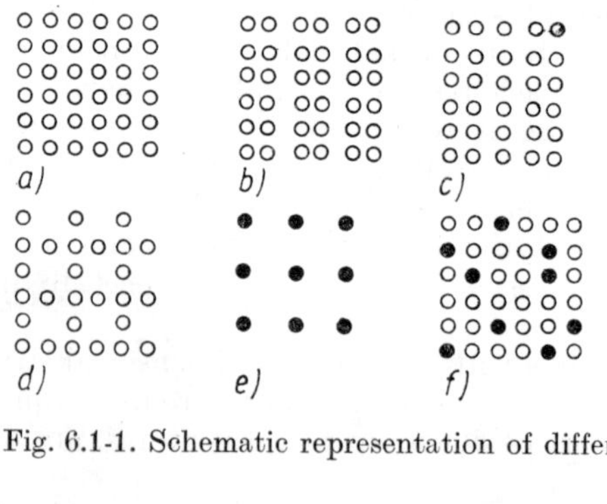

a) b) c) d) e) f)

Fig. 6.1-1. Schematic representation of different surface structures

(c) In these superstructures, a few new types of defect may appear; the defect shown in fig. c) could be denoted as a two-dimensional twin or a domain boundary.
(d) In fig. d), an incomplete superstructure of the layer is represented. The surface layer shows periodicity as before, but its occupation density is reduced in comparison with the bulk.

The next structures concern chemical deviations. Here only the simplest cases are indicated

(e) For periodical arrangements of *ad-atoms* as in fig. e) (we consider them mainly as adsorbed impurity atoms) the remarks made to figs. a) to d) are also valid for the possible configurations. As a consequence of the comparably weak interaction of the ad-atoms, there is a finite probability for them to change their positions by thermal excitation.
(f) If the covering of the surface is small or the mutual interaction of the ad-atoms is weak in comparison with the thermal excitation energy, they can be distributed statistically over the possible stable sites of the surface (fig. f)).
(g) The ideal structure of the surface may be disturbed by kinks and corners within it.

6.1.2 Surface States

The many possible surface shapes are important since the electronic structure of the surface more or less depends upon it. Since at present the experimental determination of the surface structure is in its infancy and, furthermore, as the calculation of the exact behaviour of the potential near the surface is a very complicated problem even if the details of the surface structure are known, it is not sensible to make theoretical predictions on the electronic structure too detailed, without strong connection to precise experimental data. Therefore, theoretical papers concerning the electronic structure of surfaces mainly restrict themselves to general statements comparably independent of details of geometrical structure, like the appearance of *surface states*.

As surface states, we consider states localized at the surface with respect to its normal. In the bulk direction, they decrease exponentially. Within the surface they may be extended or localized as a function of the given surface structure. Calculating them, the electronic structure of the bulk is considered to be known. The reason for the appearance of surface states lies in the changed boundary condition at the surface. As discussed in connection with the Bloch condition (1.2-5), whilst going from cell to cell the wave function for energies within the allowed band is multiplied by a factor of modulus one, it can increase exponentially within the forbidden energy region. The solution increasing in the direction from inside to the surface is physically meaningful, if it may be fitted continuously and with continuous tangents to the solution outside the crystal decreasing exponentially into the vacuum (of course the energy of the surface state has to be smaller than the potential outside the crystal, since in that case the electron is able to escape from the metal). The situation is represented in fig. 6.1-2 for a one-dimensional case. The lower fig. shows schematically the behaviour of the potential within the crystal with an assumed, in our model, sudden jump to its external value. The energy in the forbidden region induces the exponential increase of the envelope of the otherwise oscillating wave function. At the surface, the internal and external solutions are joined together with equal logarithmic derivatives ψ'/ψ (of course this condition can be fulfilled only at the exact energy of the surface state!). If we shift the surface (e.g. the position of the jump) as in fig. 6.1-2 by nearly half a lattice constant, the logarithmic derivatives inside and outside have different signs and, therefore, in the total forbidden gap a surface state cannot exist, since even for the most general potential the logarithmic derivative ψ'/ψ cannot change its sign within a forbidden gap (Lehmann 73). This illustrates that the existence of surface states depends very sensitively on the real shape of the potential, which has to be calculated in a self-consistent manner.

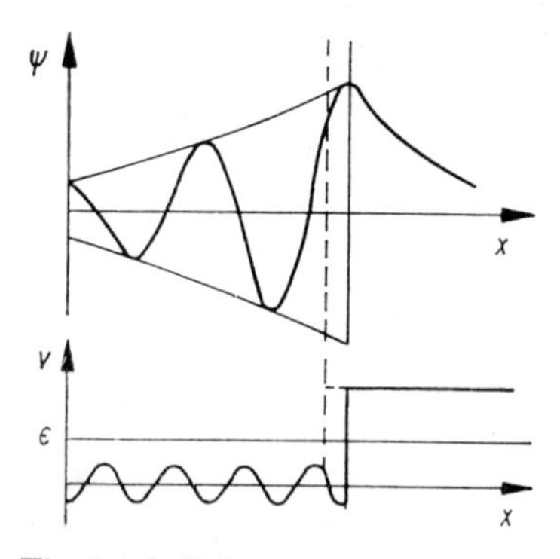

Fig. 6.1-2. Behaviour of potential $v(x)$ and wave function $\psi(x)$ of surface states (schematically). In the case of the potential step at the broken line the logarithmic derivative ψ'/ψ is positive at the surface, therefore, no surface state can exist

The exponential increase of the wave function can be described formally assuming $\boldsymbol{k}$ to be a complex quantity in (1.2-5). In this sense, we also use the term of "*complex band structures*" in treating surface states, i.e. calculation of the dipsersion relation $\varepsilon(\boldsymbol{k})$ and the wave function for complex values of $\boldsymbol{k}$.

Because of the periodicity within the surface plane (assumed to be the xy-pane), in the three-dimensional case k_x and k_y remain eigenvalues of the translation operator. The possible energies $\varepsilon(k_x, k_y)$ also depend on them and so form a (two-dimensional) band of surface states. For fixed values of k_x and k_y, they are situated in the energy region forbidden in the bulk with respect to k_z.

In the case of superstructures within the surface, several surface bands may appear.

In materials with strongly varying crystal potential (e.g. transition metals), the fit of the wave functions at the surface produces large difficulties since, because the functions $\mathfrak{a}$ in (1.2-5′) are not even approximately constant, the inner wave functions couple with several plane waves decreasing exponentially in the external region and discriminating with respect to k_x and k_y by vectors of the two-dimensional reciprocal lattice.

In semiconductors, surface states are important because of surface space charges induced by them, the creation of trapping centres and of recombination centres, short circuit channels, and other undesired effects. In metals, they are important in connection with chemical phenomena at surfaces like corrosion, catalysis, and also for the surface energy.

The influence of the changed boundary conditions on states within a band is small, but it may be observed in the energetic distribution (local DOS) of the band electrons near the surface.

6.1.3 Thin Layers

As regards electronic structure of thin layers, we should discuss several aspects.

(a) The influence of surface states increases with decreasing layer thickness, since the number of those states becomes more important with respect to the bulk number of states.

(b) The dimensionality effect means that for the motion of electrons normal to the layer only discrete eigenvalues appear. Figure 6.1-3 gives the behaviour of the DOS for a simple layer model described by a constant potential with infinite potential walls. The constant contributions to the DOS correspond to the two-dimensional motion along the potential wells. A jump appears always, if at the corresponding energy the next excited state for the motion normal to the wall appears. For very thin layers the jump in the DOS may be observed in a number of measurable quantities, if their dependence on the thickness of the layer is considered.[1])

(c) Thin films may change their symmetry in comparison with the bulk material. In each case, changes of the lattice constants, work function, surface energy, and similar quantities are also to be expected.

(d) Macroscopic deviations from the ideal surface structure are also noticeable in thin layers. In particular, they are important for evaluation of transport properties. Usually, they are described in a semi-microscopic empirical way by means of the coefficient p of specular reflection of an electron incident on the surface ($p = 0$: diffuse reflection, $p = 1$: specular reflection).

[1]) The measurement of the one-electron tunnel current in very thin layers shows, for example, strong periodic fluctuations as a function of layer thickness.

Linear methods of band structure calculation, such as LAPW (see [2.6.3]), are used with increasing success for the calculation of systems of layers (up to about 20 layers). Also possible are statements on measurable quantities such as work function, relaxation effects, and the interaction of ad-atoms on the surface (e.g. (001) surface on W: Posternak et al. 82; (001) surface of ferromagnetic Ni: Jepsen et al. 82).

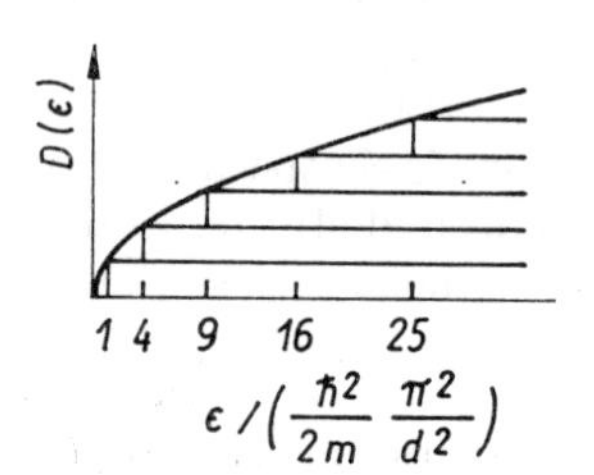

Fig. 6.1-3. Density of states $D(\varepsilon)$ in a layer model with infinite walls, thickness d. The thick line represents the DOS of free electrons

6.2 Ideal Surfaces of Metals

In this section the behaviour of the potential at metal surfaces and the physical problems connected with this will be discussed for the simplest case — simple metals (see also Paasch and Hietschold 83).

6.2.1 Formulation of the Problem

In analogy to the band structure calculations for the bulk, also here the Kohn-Sham equation of the density functional theory (1.2-1) and (1.3-7, 11, and 13) is the starting point:

$$\left(-\frac{\hbar^2}{2m}\frac{\partial^2}{\partial \boldsymbol{r}^2}+v(\boldsymbol{r})\right)\psi_n(\boldsymbol{r})=\psi_n(\boldsymbol{r})\,\varepsilon_n,$$

$$v(\boldsymbol{r})=v_{\text{ext}}(\boldsymbol{r})+\int \mathrm{d}^3r'\,\frac{e^2 n(\boldsymbol{r})}{|\boldsymbol{r}-\boldsymbol{r}'|}+v_{\text{xc}}\big(n(\boldsymbol{r})\big); \tag{1}$$

$$n(\boldsymbol{r})=\sum_n |\psi_n(\boldsymbol{r})^2|.$$

$v_{\text{ext}}(\boldsymbol{r})$ is the ionic potential, the second term is the Hartree potential of the conduction electrons, and the last term is the exchange and correlation potential. As distinct from the usual band structure calculations, the application of (1) to metal surfaces produces the following problems: perpendicular to the surface the three-dimensional translational symmetry is broken, and the electron gas is strongly inhomogeneous in the surface region. Therefore, simple model systems were first studied, in preference. In the simplest approximation, the discrete lattice structure is neglected. Thus an inhomogeneous semi-infinite electron gas remains. However, for higher electron densities, the lattice ions must be taken into account.

6.2.2 Semi-Infinite Electron Gas

Within this model, the ionic charge (in the half space $z < 0$) is thought to be uniformly smeared, i.e. $n_i(z) = \bar{n}\theta(-z)$. This background distribution according to (1) causes an electron density $n(z)$ with $n(-\infty) = \bar{n}$ and $n(+\infty) = 0$ (see Fig. 6.2-1). The total charge density is given by

$$en_{\mathrm{D}}(z) = e\big(n(z) - \bar{n}\theta(-z)\big), \tag{2}$$

from which, via the Poisson equation $-\varepsilon_0 \Delta u_{\mathrm{D}}(z) = en_{\mathrm{D}}(z)$, the potential $v_{\mathrm{D}}(z) = eu_{\mathrm{D}}(z)$ results. This potential $v_{\mathrm{D}}(z)$ occurs in (1) instead of the first two terms of $v(\boldsymbol{r})$, and hence

$$v(z) = v_{\mathrm{D}}(z) + v_{\mathrm{xc}}\big(n(z)\big) \tag{3}$$

with $v(\infty) = v_{\mathrm{D}}(\infty)$ and $v(-\infty) = v_{\mathrm{D}}(-\infty) + v_{\mathrm{xc}}(\bar{n})$.

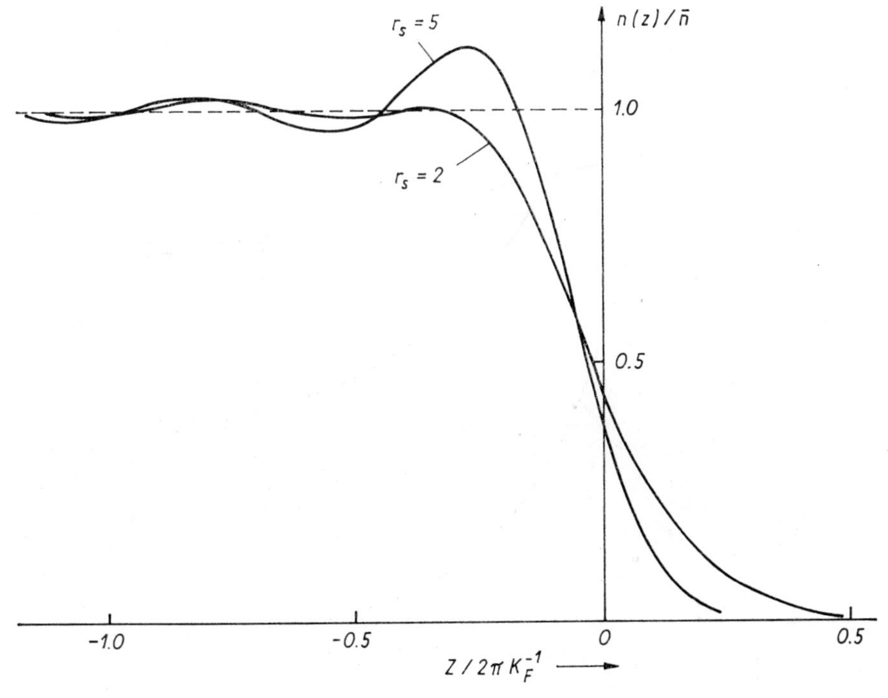

Fig. 6.2-1. Electron density $n(z)$ of a semi-infinite electron gas with r_s as dimensionless density parameter

Because electrons spill over from the half space $z < 0$, a dipole layer arises near the surface at $z = 0$. This is characterized by a potential difference (dipole barrier)

$$D = v_{\mathrm{D}}(\infty) - v_{\mathrm{D}}(-\infty) = 4\pi\epsilon^2 \int_{-\infty}^{\infty} \mathrm{d}z z n_{\mathrm{D}}(z)\,. \tag{4}$$

The work function Φ is the minimum energy necessary to remove one electron from the solid at $T = 0$, and is therefore given by

$$\Phi = v(\infty) - \mu\,. \tag{5}$$

Here, the electrochemical potential μ is the highest energy of the one-particle states, occupied according to (1). With (3) and (4) it can be written

$$\Phi = D - \zeta\,, \quad \zeta = \mu - v_{\mathrm{D}}(-\infty)\,, \tag{6}$$

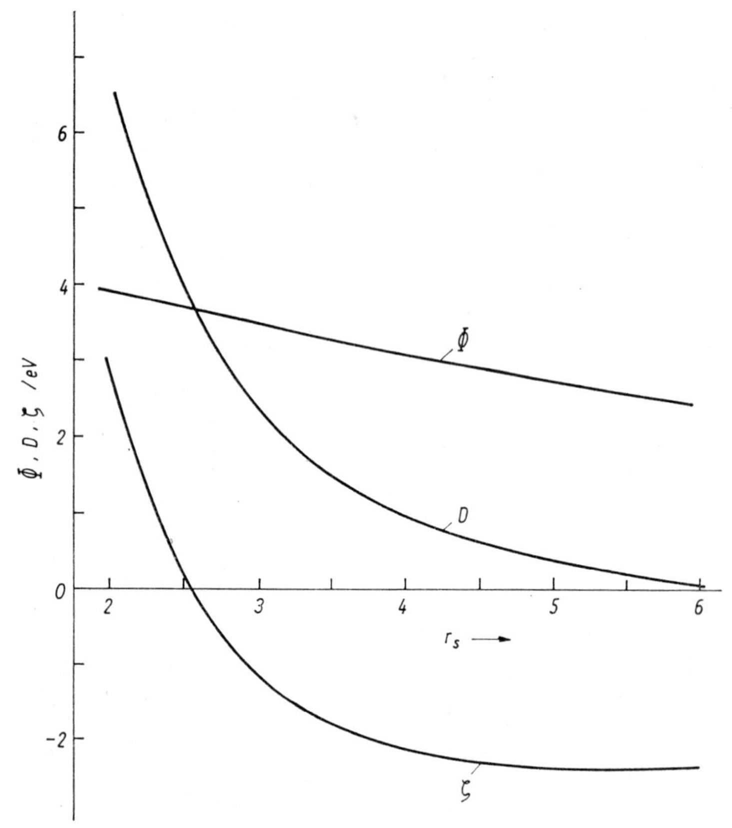

Fig. 6.2-2. a) Work function Φ, dipole barrier D, and chemical potential ζ of the semi-infinite electron gas as a function of r_s

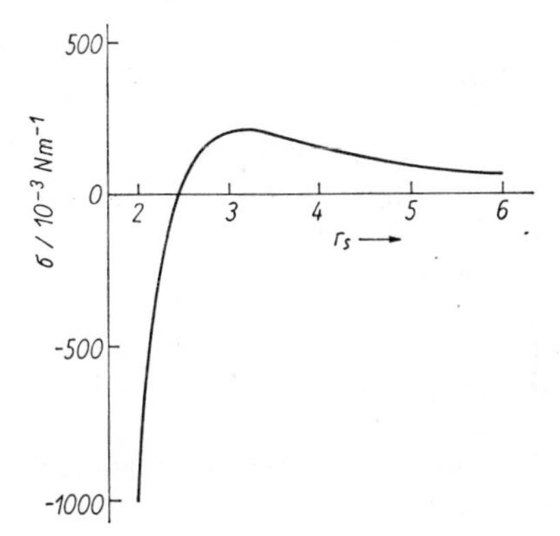

Fig. 6.2-2.b) Surface energy σ of the semi-infinite electron gas

where ζ is the chemical potential with respect to the electrostatic potential $v_{\mathrm{D}}(-\infty)$ in the bulk. With (1) it results that $\zeta = \hbar^2 k_{\mathrm{F}}^2/2m + v(-\infty)$, and thus $\zeta = \hbar^2 k_{\mathrm{F}}^2/2m + v_{\mathrm{xc}}(\bar{n})$ with $\bar{n} = k_{\mathrm{F}}^3/2\pi^2$ according to (1.1-5).

For a given background density $\bar{n} = 3/4\pi a_0^3 r_s^3$ and exchange and correlation potential $v_{\mathrm{xc}}(\bar{n})$, given within or beyond LDA, the self-consistent solution of (1) delivers as well as $n(z)$, also $v_{\mathrm{D}}(z)$, $v(z)$, D, μ, Φ, and also the surface energy σ. The latter quantity determines via $E_s = A\sigma$ the energy required to create a new free surface A; cleaving a bulk specimen, two free surfaces are produced. Lang and Kohn (70) used the Wigner approximation for v_{xc} which just interpolates between the limiting cases of metallic and vanishing electron density (this should be relevant for surfaces). They obtained the results shown in Fig. 6.2-2. For small values of the density parameter r_s the quantities ζ and σ become positive and negative, respectively; this is physically unreasonable and shows the limits of the model. The values of D are also too large for small r_s compared with semi-empirically estimated data. The agreement of Φ with the experimental values of the most simple metals only within a range of 0.5 eV is a consequence of the ("accidental") compensation of errors of its parts D and ζ.

6.2.3. Variational Treatment

Within density functional theory (1.3-10), a simplified treatment (but capable of elaboration) is possible by means of the variational ansatz for the density

$$n(z) = \bar{n}\left(\left(1 - \frac{1}{2}\,\mathrm{e}^{\beta z}\right)\theta(-z) + \frac{1}{2}\,\mathrm{e}^{-\beta z}\theta(z)\right), \tag{7}$$

where β is a parameter still to be determined (Smith 69). Thus for the density a form is assumed that describes the real physical situation. It contains only one variational parameter, but the simplicity of (7) allows us to go beyond

LDA using the gradient expansion method (known from the many-body theory of the inhomogeneous electron gas) and including LFC's of the dielectric function (cf. [5.2.3]).

The parameter β as a function of r_s is determined by calculating the surface energy σ following from (7), and by minimizing this $\sigma(\beta, r_s)$ with respect to β. Using LFC's known from the literature yields curves $\beta(r_s)$ which show only a weak dependence of r_s and are close to $1/a_0$. The resulting curves $\sigma(r_s)$ and $D(r_s)$ agree qualitatively well with those of Lang and Kohn (70). The drawback of the variational treatment is that the electron density is not determined self-consistently, and that the ansatz (7) does not contain the Friedel oscillations (see [5.5.2]), which are more distinct especially for larger r_s values.

The advantage is the possibility already mentioned of going beyond LDA and taking into account the ion potentials. The latter will be described in the following.

6.2.4 Ion Lattice Model

In this case, the ions are described by model potentials of the form (2.2-10). Here, due to Ashcroft, $A_l = 0$ and $R_l = r_c$ is assumed, r_c is the only model parameter of the ion potential. The top layer is allowed to shift as a whole by δ in comparison with its ideal bulk position, and so σ depends on the variational parameters β and δ describing the electron and the lattice relaxations, respectively. Their values result from the minimizing of $\sigma(\beta, \delta; r_s)$. The dipole barrier $D = 4\pi e^2 \bar{n}(\beta^{-2} - d\delta)$ apart from the above-mentioned electron contribution (first term), now contains an addiational ion contribution (second term with d being the layer spacing of the bulk), which plausibly decreases the dipole moment and hence also the dipole barrier D if the top layer is shifted outward ($\delta > 0$).

Table 6.2-1 shows the theoretical results for β, δ, σ and D with Al and Na, where the LFC of Singwi et al. (70) has been used. Altogether it was possible by means of the variational treatment to achieve a quite good agreement of the calculated values of σ, D, Φ and the anisotropy of these quantities with experimental data for a large number of simple metals (Paasch and Hietschold

Table 6.2-1. Characteristic parameters of Al-(111)-*bcc* and Na-(110)-*bcc* surfaces.
Experimental and semi-empirical values, respectively, are given in parenthesis.

	r_c/a_0	r_s	δ/a_0	β/a_0^{-1}	$\sigma/10^{-3}$ Nm^{-1}	D/eV
Al	1.109	2.073	0.053	1.29	862 (860)	3.36 (3.4)
Na	1.711	3.931	0.119	0.90	221 (220)	0.76 (0.7)

77 and references quoted there). It thus becomes evident that σ increases with decreasing r_s and increasing valence, which is in agreement with experimental data for one- and two-valent metals (exceptions are Be, Zn, Hg, for which the Ashcroft potential is too simple an approximation). For three- and four-valent metals values of σ often too large result. Reasons for that might be: too simple a model potential (Sn, Ga); repulsion of the d-electrons in closed shells; inhomogeneity of the electron gas perpendicular to the z-axis (especially for metals with small r_s values).

6.2.5. Recent Developments

Recent developments in the ab-initio electron theory of surfaces concern e.g.

- the oscillatory multilayer relaxation of the Al (110) surface (Ho and Bohnen 85) and other metal surfaces (Inglesfield 85, Chen et al. 86, Jiang et al. 86, Sokolov et al. 86, Allan and Lannoo 88, Luo and Legrand 88), also semiconductor (Qian et al. 87) and insulator (de Wette et al. 85) surfaces show this phenomenon, multilayer reconstructions and vibrational properties of the W(001) surface (Fu and Freeman 88),
- the electronic screening at the surface of an sp-bonded metal and application to static phenomena (lattice relaxation at Al (110)) and dynamic processes (damping of a vibration of an adsorbed molecule due to decay into electron-hole pairs, excitation of collective modes in an overlayer), see Eguiluz (87),
- the missing-row reconstruction on *fcc* (110) transition-metal surfaces (Ho and Bohnen 87),
- the adatom induced 3×3-reconstruction (Takayanagi reconstruction) of the Ge (111) surface (Payne 87),
- the rippled NiAl (110) surface (Lee et al. 87),
- the increase of the magnetic moment of an atom in the top layer of the Ni(001) surface in comparison with the moment of a bulk atom (Jepsen et al. 82) as an example of the pecularities of the surface magnetism in addition to those mentioned at the end of [3.2],
- the image potential, the position of the image plane, and the image-potential states of simple metal surfaces, studied by using the semi-infinite jellium model and an appropriate non-LDA for the XC-potential (Ossicini et al. 86, Gies 86),
- the jellium surface in a strong static electric field (Gies and Gerhardts 86, Ossicini et al. 87),
- there are attempts to discuss surface properties in terms of an appropriately defined surface stress tensor (Needs 87, Vanderbilt 87, Payne 87, Dodson 88); a surface stress theorem as a dyadically generalized surface virial theorem and force-sum rules at unrelaxed surfaces, which give a general frame for the above mentioned multilayer relaxations, have been derived (Kaschner and Ziesche 88a, b).

This list of problems and corresponding references is by no means complete. Core level shifts, configuration relaxations, surface states, surface plasmons, surface phonons and other collective surface modes, physi- and chemisorption, surface properties of alloys e.g. have to be added. Single adatoms are statically (geometry) and dynamically (vibrational frequencies, barriers for lateral diffusion and vertical penetration) studied, the direct and indirect interaction between adatoms (the latter via the substrate) are investigated. Dissociative adsorption means the spontaneous breaking of bonds of molecules approaching a metal surface. Surfaces, i.e. their chemical and physical structure as well as their dynamics are intensively investigated by means of scattering processes, new experimental tools to study surfaces are the scanning tunnel microscope and the atomic force microscope. Each of these methods needs for a reliable interpretation of the measured data a corresponding electron theory.

6.3 Systems with Low-Dimensional Structure

As systems with low-dimensional structure, we denote solids where the strength of bonds in a certain direction or within a plane dominates, called quasi-one-dimensional and quasi-two-dimensional, respectively. Of special interest is the fact that the electrical conductivity of such systems may reach the order of the conductivity in metals without containing typical metallic elements. Therefore, they are also called as synthetic metals.

6.3.1 Quasi-Two-Dimensional Conductors

Among the quasi-two-dimensional systems, the following are of special interest:

- dichalcogenides of transition metals (e.g. MoS_2),
- layered halogenides as PbJ_2, $CuCl_2$, or Ag_2F,
- semiconducting chalcogenides as GaS, GaSe, As_2Se_3,
- the intercalation compounds of graphite, and
- the recently discovered high-temperature superconductors (Bednorz, Mueller 86) with perovskite-like structures ($La_{1,8}Sr_{0,2}CuO_4$, $YBa_2Cu_3O_{7-\delta}$, see [5.1.3]).

As an example of layered structures, we shall discuss graphite in more detail. The threefold coordination of carbon atoms represented in fig. 6.3-1 is characteristic for such compounds leading to an open cyclic structure. The states presented schematically in fig. 6.3-1 are responsible for this structure. They are generated by hybridization of one atomic $2s$-state with two atomic $2p$-states and characterized by a club-shaped electron distribution forming tightly bound homopolar bonds with the neighbouring carbon atoms. Each of the neighbouring carbon atoms contributes one electron to the bond. The residual of the four valence electrons of each carbon atom occupies an atomic p-state with an extension perpendicular to the layer plane. The overlap of the wave functions perpendicular to the layer and the bonds in this direction are weak, so that the graphite planes can easily slide over each other (strictly speaking only van-der-Waals bonds exist between separate layers). In the three-dimensionally structured (metastable) modification of carbon — diamond — the $2s$-electron hybridizes with all three $2p$-electrons. In this way, a four-fold tetrahedral (spatial) coordination represented in fig. 6.3-1 c) originates leading to the cubic diamond lattice. Contrary to graphite, in diamond all four valence electrons participate in bonding which induces the extraordinary hardness of diamond.

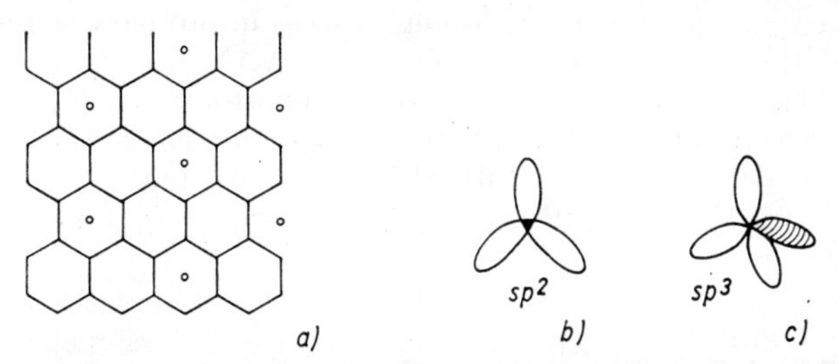

Fig. 6.3-1. a) Layer lattice of graphite. The circles denote the position of the Li atoms in the intercalation compound LiC_6.
b) Electron distribution of the carbon atom in the configuration sp^2 (graphite). The bonds are situated within the plane of the drawing.
c) Three-dimensional configuration sp^3 (diamond)

Because of the small overlap of the p-wave functions in graphite (which because of their symmetry properties with respect to the layer plane are also called π-states), their levels split in the solid much less than the σ-bands. The bonding and antibonding π-band overlap somewhat, so that the two π-electrons in an elementary cell (one electron per carbon atom) do not fill the bonding band completely. Approximately 10^{-4} electrons per carbon atom flow into the antibonding band, leaving this number of holes in the bonding band. Therefore, pure graphite must be characterized as a semi-metal with an electrical conductivity of $\sigma = 2.5 \times 10^4$ S/cm at room temperature. (5.1-20) cannot be used immediately for the calculation of the conductivity because the energy bands in graphite near the Fermi energy have a linear dispersion relation $\varepsilon = \hbar v_F k$. Therefore, we obtain for the conductivity in the two-dimensional case (c denotes the lattice constant normal to the layer plane)

$$\sigma = \frac{e^2 v_F \tau}{\hbar \sqrt{2\pi c}} \sqrt{n}. \quad (1)$$

If because of the small charge carrier concentration n of graphite in comparison with metals (where nearly each of the z valence electrons of an atom contributes to the conductivity), the conductivity of graphite is smaller than that of copper (5.9×10^5 S/cm) only by a factor 20, it is caused by the high mobility of charge carriers in graphite. The reason is the high hardness of a graphite layer which is induced by the homopolar bonds and which results in small amplitudes of lattice oscillations (a high Debye temperature) and in a smaller scattering probability τ^{-1} of electrons on the lattice vibrations. If one succeeds in increasing n considerably without reducing τ, a conductivity comparable with that of the best metals may be obtained. This is possible by introducing atomic or molecular layers between the graphite layers which deliver electrons (donors) and accept electrons (acceptors), respectively. This process is called *intercalation*. Through this, the graphite structure within a layer is perturbed only slightly by the ordered arrangement of the interlayers and, therefore, as opposed to the doping process in ordinary semiconductors, the relaxation time is reduced only insignificantly. As donors, alkaline metals may be used. The regular arrangement of lithium atoms in LiC_6 is indicated in fig. 6.3-1a); acceptors are strongly oxidizing compounds such as AsF_5 or SbF_5. By intercalating graphite with SbF_5, a conductivity at room temperature of approximately 4.4×10^5 S/cm (nearly 70% of that of copper) is obtained.[1]

6.3.2 Quasi-One-Dimensional Conductors

A typical selection of quasi-one-dimensional structures and of attainable conductivities is given in fig. 6.3-1. Different groups can be distinguished

(1) As the representative of the group of charge transfer complex salts, we mention tetrathiofulvalene-tetracyanoquinodimethane (TTF-TCNQ).[2] The solid is formed by ordered stacking of planar TTF and TCNQ molecules, respectively. The conductivity is caused by a charge transition between

[1] Values of 6×10^5 S/cm and higher for conductivities (already published in 1977 by Vogel) could not be confirmed (see Wu et al. 81).

[2] TCNQ molecule

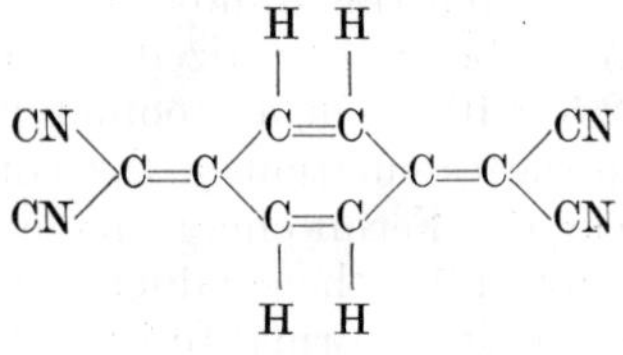

```
                H  H
                |  |
    CN        C==C         CN
      \      /    \       /
       C==C        C==C
      /      \    /       \
    CN        C==C         CN
                |  |
                H  H
```

TTF molecule

```
  H     S                S     H
   \   / \              / \   /
    C     \            /     C
    ||      C==C             ||
    C     /            \     C
   /   \ /              \ /   \
  H     S                S     H
```

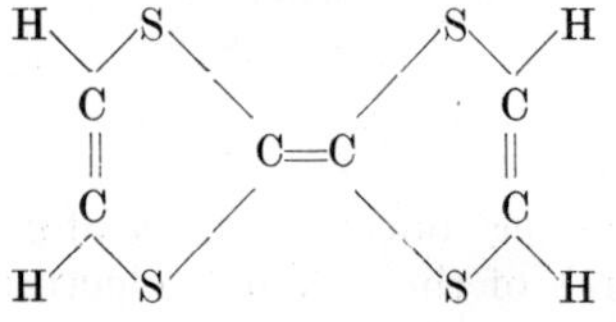

Table 6.3-1. Quasi-one-dimensional materials (SC: superconductor)
Tr: room temperature
GIC: graphite intercalation compound

compound	peculiarity	$\sigma(T_{zi})$ / $(\Omega\ \text{cm})^{-1}$	σ_{max} / $(\Omega\ \text{cm})^{-1}$
1) TTF-TCNQ	Peierls transition	900	10^5 (at 54 K)
$(TMTSF)_2ClO_4$	$T_c = 1.3$ K SC		
$(TMTSF)_2PF_6$	$T_c = 0.9$ K SC (under pressure)		10^5 (at 15 K)
2) KCP $K_2Pt(CN)_4Br_{0,2} \cdot H_2O$	Peierls transition	300	
3) $Hg_{2,86}AsF_6$	SC		
4) $(SN)_x$	$T_c = 0.33$ K	2000	
$(SNBr_{0,4})_x$	$T_c = 0.35$ K 3D-SC	40000	
5) $(CH)_x \cdot SbF_5$	Peierls transition	3000 (stretched)	
6) $(=C=)_x$	α-Carbyne	$10^{-6} \ldots 10^{-3}$	
$(-C{\equiv}C-)_x$	β-Carbyne		
7) TaS_3	Peierls transition	≈ 1000	
$NbSe_3$			
8) graphite	semi-metal	$2.5 \cdot 10^4$	
GIC (SbF_5)	two-dimensional	$4.4 \cdot 10^5$	
GIC (AsF_5)		$4.1 \cdot 10^5$	
GIC (C_8K)	$T_c = 0.2$ K SC		
9) copper	metal	$5.9 \cdot 10^5$	
aluminium	metal three-dimensional	$3.9 \cdot 10^5$	

both kinds of stacks. It increases with decreasing temperature to values of approximately 10^5 S/cm, below 54 K the complex has an insulating ground state. TMTSF (tetramethylselenofulvalene) also forms stacks of planar molecules. A very interesting feature is the transition of the alloys of TMTSF with ClO_4-ions into the superconducting state with a critical temperature T_c of 1.3 K at normal pressure (Bechgaard et al. 81).

(2) Potassium cyanoplatinate (KCP) has been known as a compound since 1842. Its conductivity is caused by the chain-like arrangement of the platinum atoms and by doping with bromine.

(3) $Hg_{2,86}AsF_6$ also exhibits a chain-like structure and superconductivity at low temperatures.

(4) The polymer $(SN)_x$ has been known since 1910, but its superconducting properties were not discovered before the beginning of the seventies. It was very surprising since polymeric superconductors were not known before. It has now been clarified that the well known electron-phonon interaction is responsible for superconductivity in this polymer, as in metallic systems.

The interaction between the polymer chains is quite strong so that an anisotropy of conductivity of the order of only 5 is observed.

(5) The electrical conductivity of the well insulating polyacetylene $(CH)_x$ as the representative of conducting conjugated polymers may be increased by doping by more than 10 orders of magnitude up to a value of 8×10^4 S/cm (Schimmel et al. 88).

(6) The one-dimensional modifications of carbon (the α- and β-carbyne) are also remarkable. They consist of carbon chains with double bonds and alternating single and triple bonds, respectively. Before now, carbyne was little studied because samples are only producible under extreme conditions and with small dimensions, below 1 μm.

(7) Trichalcogenides of transition metals also form chain-like structures (e.g. $NbSe_3$).

(8) Further examples are lead-phthalocyanine consisting similarly of stacks of molecules and showing bistable conductivity states (Hamann et al. 78) and the so-called blue bronze ($K_{0.3}MoO_3$).

6.4 Peculiarities of Low-Dimensional Systems

The peculiarities of low-dimensional systems become evident in their electronic structure and in their reaction on external perturbations. Therefore, in fig. 6.4-1 the Fermi body, the DOS, and the dielectric susceptibility χ are represented for one-, two-, and three-dimensional systems. In each case, we assume a simple quadratic dispersion relation $\varepsilon \sim k^2$. In the three-dimensional case the Fermi body is a sphere, in the two-dimensional case, because of the missing interaction between the layers and the missing dispersion in the direction normal to the layers, it is a cylinder.[1]) In the one-dimensional case, the

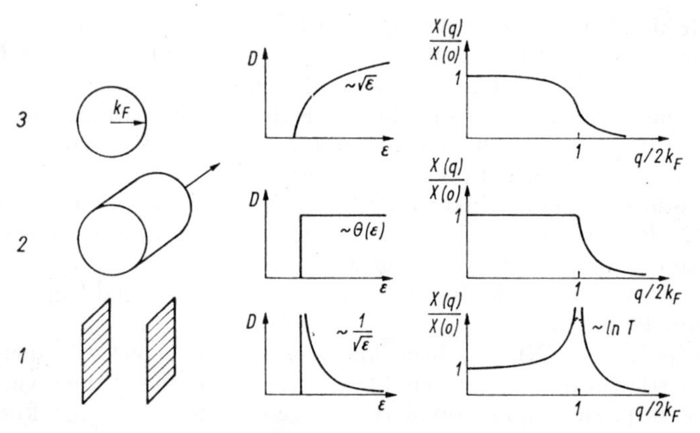

Fig. 6.4-1. Fermi body, DOS, and susceptibility $\chi(q)$ of three-, two-, and one-dimensional systems

energy depends on the momentum in the chain direction only. Hence the Fermi surface consists of a pair of planes. The DOS (1.2-24) results from the number of states between surfaces of constant energy, and therefore it yields as a function of energy singularities of different strengths at the band edges which are displayed in fig. 6.4-1.

According to (5.2-17), the susceptibility $\chi(\boldsymbol{q})$ characterizes the reaction of the system (in the form of a change of density δn_{ind}) on a time-independent spatially periodic external perturbation $\delta V_q \sim \operatorname{Re} \exp(iqx)$ with wavelength $\lambda = 2\pi/q$,

$$\chi(\boldsymbol{q}) = \frac{\delta n_{\text{ind}}}{\delta V_{\boldsymbol{q}}} \sim \sum_{\boldsymbol{k}} \frac{f\big(\varepsilon(\boldsymbol{k})/k_{\mathrm{B}}T\big)\big[1 - f\big(\varepsilon(\boldsymbol{k} - \boldsymbol{q})/k_{\mathrm{B}}T\big)\big]}{\varepsilon(\boldsymbol{k}) - \varepsilon(\boldsymbol{k} - \boldsymbol{q})}. \tag{1}$$

The energy denominator shows that an electron in the state $\varepsilon(\boldsymbol{k})$ may be excited more easily, the smaller the difference from the energy $\varepsilon(\boldsymbol{k} - \boldsymbol{q})$ of a state excited by the momentum $\boldsymbol{q}$ of the external field. The product of Fermi distribution functions of type $f(\boldsymbol{k})\,[1 - f(\boldsymbol{k} - \boldsymbol{q})]$ in the denominator guarantees that only occupied states $\boldsymbol{k}$ are excited into unoccupied ones $(\boldsymbol{k} - \boldsymbol{q})$. The terms in (1) diverge for all states $\boldsymbol{k}$ where the denominator vanishes,

$$\varepsilon(\boldsymbol{k}) - \varepsilon(\boldsymbol{k} - \boldsymbol{q}) \sim -\boldsymbol{q}(\boldsymbol{q} - 2\boldsymbol{k}) = 0\,. \tag{2}$$

(2) is the equation of a plane normal to $\boldsymbol{q}$ with distance $k = q/2$ from the origin. Performing the integration in (1) over $\boldsymbol{k}$, the singularities of type $1/(q - 2k)$ in (1) cancel except for the case where the plane (2) touches the region of occupied states, meaning that it touches the Fermi body. In other words, a singullarity in (1) appears if q equals a diameter of the Fermi body. Then the maximum of occupied states $\boldsymbol{k}$ is excited into unoccupied states $(\boldsymbol{k} - \boldsymbol{q})$ with the same energy, by momentum $-\boldsymbol{q}$, meaning that they are in geometrical resonance with the external excitation. In the three-dimensional case, these states occupy a segment of the Fermi sphere and the evaluation of the integral in (1) leads to a singularity of type $(q - 2k_{\mathrm{F}}) \ln (q - 2k_{\mathrm{F}})$ marked in fig. 6.4-1.

In the two-dimensional case, many more electrons fulfil the resonance condition which causes the stronger step-like singularity in χ.

In the one-dimensional case, all electrons on the Fermi surface fulfil condition (2), and the logarithmic singularity shown in fig. 6.4-1 results. Because of the smearing out of the Fermi distribution functions for non-zero temperatures T, in (1) a sharp maximum with a height $\sim \ln T$ appears.

For low-dimensional systems, the following conclusions can be drawn:

— singularities in physical quantities increase with decreasing dimensionality,

— the reaction of the system on an external perturbation with twice the Fermi momentum also increases (since more or all electrons at the Fermi energy

[1]) A weak coupling present in real systems leads to a weak dispersion in axial direction. This causes the cross-section of the cylinders to fluctuate (warped cylinders). In quasi-one-dimensional systems the surfaces of constant energy are also wavy planes.

fulfil the geometric resonance condition), and as a consequence a very strong electron-lattice interaction at that momentum also appears,
- therefore, the excitation spectra change completely in some cases,
- the anisotropies of physical properties can obtain extreme ratios, e.g. in graphite intercalated with AsF_5 the electrical conductivity shows an anisotropy ratio of 10^6,
- in disordered one-dimensional structures all states are localized, even for small disorder (see [5.1.5]),
- the thermal and quantum mechanical fluctuations increase with decreasing dimensionality. Therefore, for short range forces (which are, for example, important for the formation of phonons) in strictly one-dimensional systems, no phase transitions at finite temperatures are possible (Landau, Lifshitz 79).

In the three-dimensional case, singularities resulting from the interaction of phonons with electrons at the Fermi energy are generally only weakly visible as Migdal-Kohn anomalies in the phonon spectra. In the one-dimensional case in KCP, we are dealing with giant Kohn anomalies. As a consequence of the interaction between the layers and chains, fluent transitions between the above ideal limits are observed. Therefore, we have *quasi-low-dimensional structures*. A weak coupling between layers or chains even justifies neglect of quantum fluctuations in these systems.

Besides the electron-phonon interaction, the Coulomb interaction is also important for the properties of low-dimensional systems. Usually in the literature the on-site exchange energy U of the Hubbard model (see Eschrig 83, section [5.2.3]) is given as a measure for the latter. The former dominates if the band width W is larger than U.

One-dimensional structures are also of interest from a pure theoretical point of view since in that case model systems with interaction often allow exact solutions.

6.5 Peierls Instability

In quasi-one-dimensional systems, the strong electron-phonon interaction for wave numbers $q = 2k_F$ induces a softening of the phonon mode with that wave number and below the Peierls temperature T_p a lattice instability occurs resulting in the form of a super-lattice with shifts $u(x)$ of the atoms from their original equilibrium positions:

$$u(x) = u_0 \cos (qx + \Phi). \tag{1}$$

This induces a perturbation for the electrons

$$V(x) = 2\Delta \cos (qx + \Phi); \quad \Delta = gu_0 \tag{2}$$

with g as the electron-phonon interaction constant. The periodic potential (2) induces the appearance of forbidden gaps of width 2Δ at $k = q/2$ represented in fig. 6.5-1. For other k-values only a correction of second order in Δ appears. A maximal energy is gained, if all energetically lowered states are occupied (hatched region in fig. 6.5-1). This is the case if the already mentioned condition

$$q = 2k_F \tag{3}$$

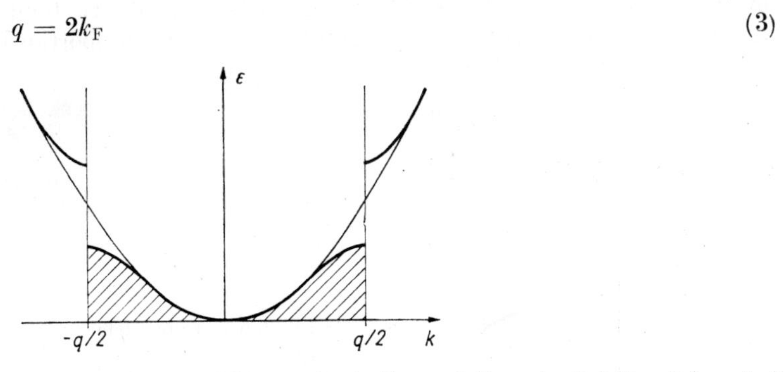

Fig. 6.5-1. One-particle energies in the periodic potential $V = 2\Delta \cos (qx)$

is fulfilled. The summation of the energy gain of all electrons yields E_e

$$E_e = \frac{1}{2} \tilde{a}\Delta^2 \left(\ln (|\Delta|/\Delta_0) - 1/2\right) \tag{4}$$

with the constants $\tilde{a}$ and Δ_0. The derivative of E_e with respect to u_0 is the driving force for the deformation. Opposite to this, the elastic energy E_l of the lattice

$$E_l = \frac{\varkappa}{2} u_0^2 \tag{5}$$

stabilizes the undisturbed configuration (elastic modulus $\varkappa$). The equilibrium value Δ^m of the energy gap follows from the minimization of the total energy in analogy to the BCS formula (5.1-8) as

$$\Delta^m = \pm\Delta_0 \, e^{-1/\lambda} . \tag{6}$$

Depending on the model used, the constant Δ_0 equals approximately the double band width $2W$. λ is the known parameter of the electron-phonon interaction (see [5.4.1]).

$$\lambda = g^2\tilde{a}/\varkappa \quad \text{with} \quad \tilde{a}^{-1} = \pi v_F \tag{6'}$$

is valid on the basis of the molecular field approximation where lattice dynamics are neglected. Better approximations bring on partially considerable effects (e.g. the change by approximately a factor 2 of the Peierls temperature T_p discussed in the following).

An interesting point in (6) concerns the singular dependence on the interaction parameter λ in total analogy to the BCS theory of superconductivity. This implies the fact that even at arbitrary small values of g because of the logarithmic dependence in (4) a minimum always exists. The temperature influence manifests itself in a smearing out of the singularity in fig. 6.4-1 ($\ln \Delta^m \to \ln T$) which reduces $\Delta^m(T)$ (fig. 6.5-2). The Peierls temperature T_p, where Δ^m vanishes, and the energy gap for $T = 0$, are proportional according to the relation

$$2\Delta^m(0) = 3.52 k_B T_p \tag{7}$$

known from BCS theory (5.1-7).

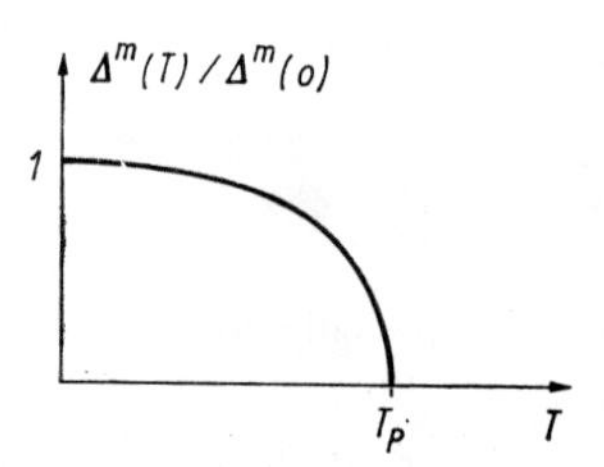

Fig. 6.5-2. Dependence of the width of the forbidden zone as a function of temperature. Above T_p the forbidden region disappears

The perturbing potential (2) induces a spatial modulation of electron density denoted as a *charge density wave* which of course is connected with the shift (1) of the atoms. For an integer or rational ratio n of the periods of superstructure and lattice, the system is denoted as *commensurate*, otherwise *incommensurate*. These relations are represented obviously in fig. 6.5-3.

The case $n = 2$ realizable in polyacetylene $(CH)_x$ (figs. 6.5-4a), b)) is known as *dimerization* and is described in chemistry by means of single and double bonds. The bonding length of a double bond (π-electrons) is reduced, its electronic density is increased in comparison to a single bond (σ-electrons; bear in mind that the π-electrons are not localized between the atoms!). Trans-polyacetylene with equal bond length (fig. 6.5-4a)) yields the ("undisturbed") dispersion relation denoted in fig. 6.5-4b) by the dashed line with crossing bands at $ka = \pi/2$. The dimerized form with interchanging single and double bonds gives the dispersion relation denoted by the solid line in fig. b) where the degeneration at $ka = \pi/2$ is lifted. From the physical point of view, the dimerization is induced by the energy gain of the charge density wave appearing during the Peierls transition. The contribution of different bonds may be characterized as follows. The σ-electrons determine the average separations of the carbon atoms (0.140 nm) and the bond angles. Induced by the π-electrons, the bond lengths of single and double bonds differ by ± 0.003 nm. Because of their interaction, the π-electrons also induce the rigidity of the molecule in the plane normal to the π-bonds.

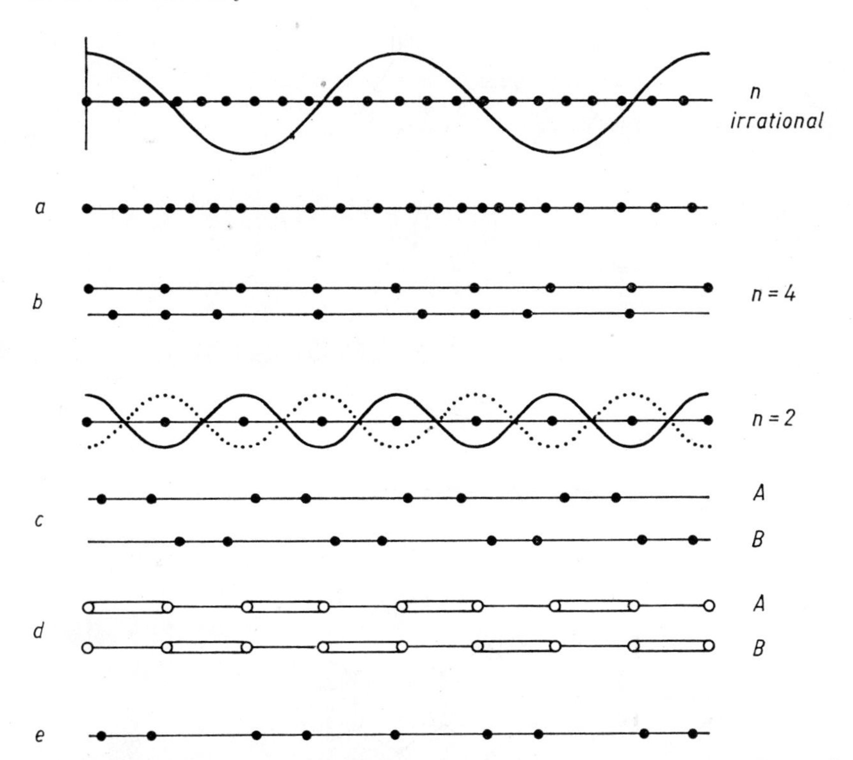

Fig. 6.5-3. Distortion of the unperturbed lattice (lattice constant a) by a charge-density wave (wavelength $2\pi/q$); unperturbed and the perturbed lattice are indicated
a) incommensurate, ($n = 2\pi/qa$ irrational)
b) commensurate, $n = 4$
c) commensurate, $n = 2$; both sensible values of Φ generate the "phases" A and B
d) representation of case c) as dimerization (alternation of single and double bonds)
e) boundary between phases A and B of fig. c) as a lattice defect

In incommensurate systems (like KCP, $NbSe_3$, TTF-TCNQ), the energy does not depend on the phase angle Φ.

As an example for the interesting transition to two-dimensional systems, we mention polyacene consisting of two tightly coupled chains (see fig. 6.5-4c)). Opposite to the crossing bands in polyacetylene, the undimerized band structure in polyacene is characterized by bands touching each other in second order. Therefore, the dimerization yields an energy gain in second order Δ^2 only, which is weaker than in eqn. (4). As a result, the arguments for the Peierls transition do not hold in their simple original form. To answer the question for the charge density wave instability, one has to take into account the finite value of λ (≈ 0.3) from (6′) and the Coulomb interaction of the electrons, which favours spin density waves.

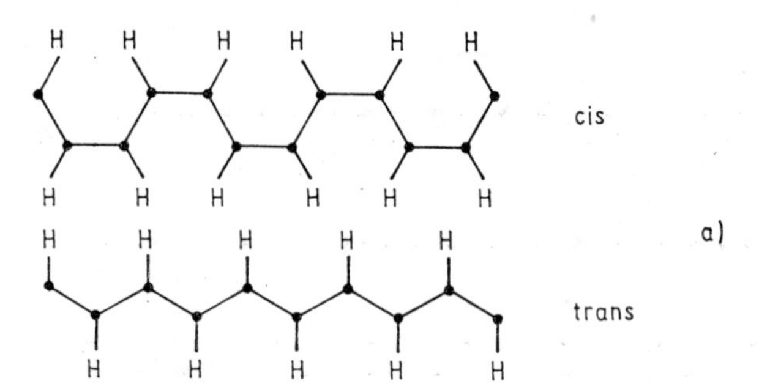
H
cis
trans
a)

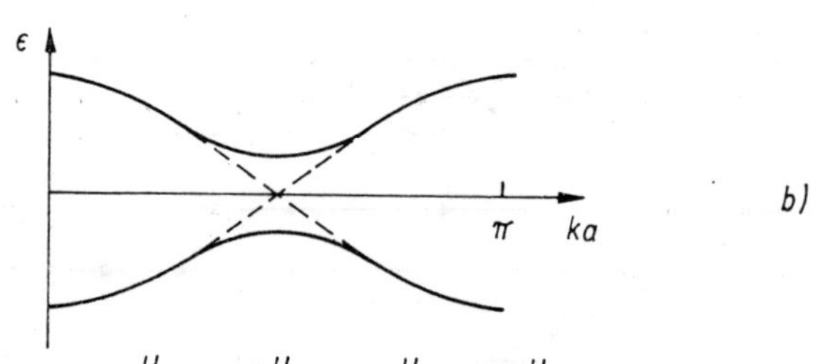
ε
π
ka
b)

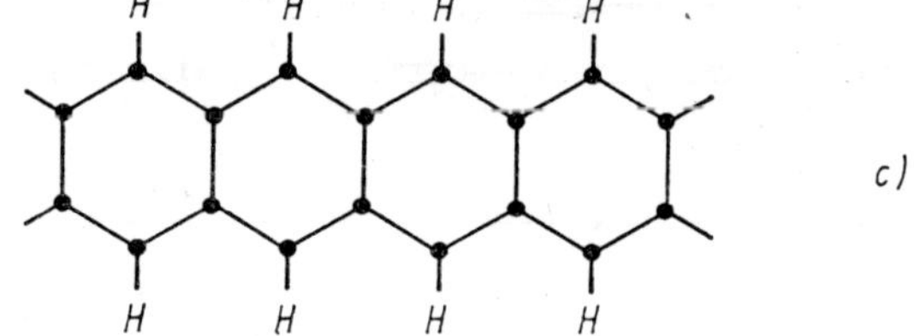
H
c)

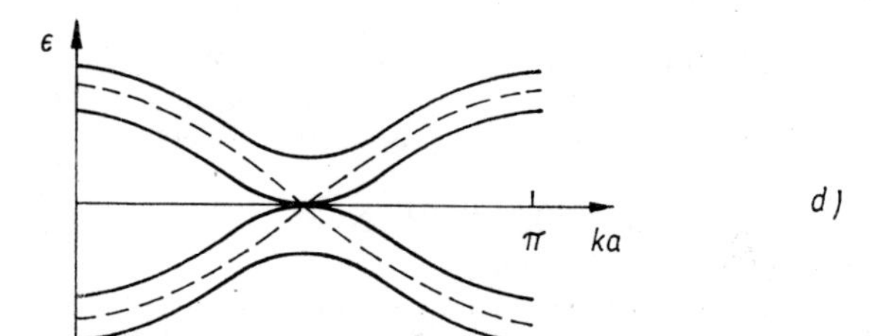
ε
π
ka
d)

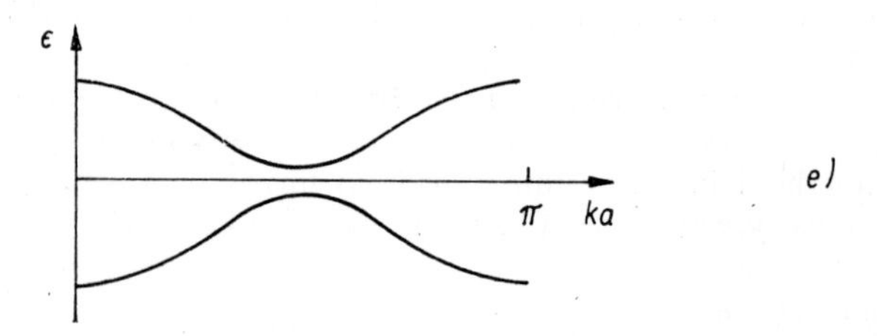
ε
π
ka
e)

6.6 Electronic Excitations in the Peierls System

The degeneration of the ground state (meaning the total equivalence of the two chains A and B in fig. 6.5-3d) shifted only by a lattice constant a) causes peculiarities of the excitation spectrum of such one-dimensional systems and denies the applicability of the picture of particle and hole excitations known from three-dimensional semiconductors. Combining the chains A and B from fig. 6.5-3c) or d) we obtain the "phase boundary" of fig. 6.5-3e) as an excited electronic state previously unknown. Of course the shift $\psi_m = (-1)^m \Delta(x_m)$ of the atoms from their unperturbed positions does not change instantaneously at the phase boundary. In $(CH)_x$ it covers according to

$$\psi_m = \Delta_0 \tanh (m/l) \tag{1}$$

with $2l \approx 14$ a region of approximately 14 C-C distances (Heeger, MacDiarmid 81). This lattice distortion is called a soliton. It is connected with a positive charge $|e| \, \Delta n_+$ induced by the shifted ions and the deformation of occupied electronic states represented schematically in fig. 6.6-1. On the other hand this modified potential causes a localized electronic state in the forbidden region separated from the continuum of states of the Peierls ground state. In the continuum model, the charge of the localized state compensates exactly the charge $|e| \, \Delta n_+$ so that the ground state with one electron occupying this state is neutral with a total spin 1/2. An occupation of the localized state with two electrons of opposite spins leads to a total negative charge $-|e|$ and a zero total spin (Su et al. 79). If the localized state is empty, we obtain a total charge $+|e|$ with zero total spin.

A detailed investigation of the instability of the extended electron and hole states as a consequence of the electron-phonon interaction has been carried out by Brazovski (80). He was able to show the stable excited states to be localized and of soliton or polaron type.

Concerning the localized excited states, it is useful to divide the Peierls systems into three classes (Brazovski, Kirova 81):

A. Incommensurate systems (KCP, $NbSe_3$, TTF-TCNQ)
 — neutral solitons (charge 0, spin 1/2)

Fig. 6.5-4. a) Cis- and trans-configuration of polyacetylene
b) dispersion relation $\varepsilon(k)$ of trans-polyacetylene (dimerized), dashed line: undimerized form with equal bond length
c) polyacene
d) dispersion relation of non-dimerized polyacene chains, for comparison the dispersion relation of two non-interacting polyacetylene chains is also given (dashed line). As a consequence of the strong interaction between both chains, their degenerate states split in the given manner.
e) dispersion relation of dimerized polyacene, the split is always of order Δ^2

B. Commensurate systems with a degenerate ground state (trans-$(CH)_x$, TaS_3)
 — neutral solitons (charge 0, spin 1/2)
 — charged solitons (charge $\pm|e|$, spin 0)
 — polarons (charge $\pm|e|$, spin 1/2)

C. Commensurate systems with non-degenerate ground state (cis-$(CH)_x$, polyparaphenylene, polypyrrole, polythiophene)
 — polarons (charge $\pm|e|$, spin 1/2)
 — bipolarons (charge 0, $\pm 2|e|$, spin 0).

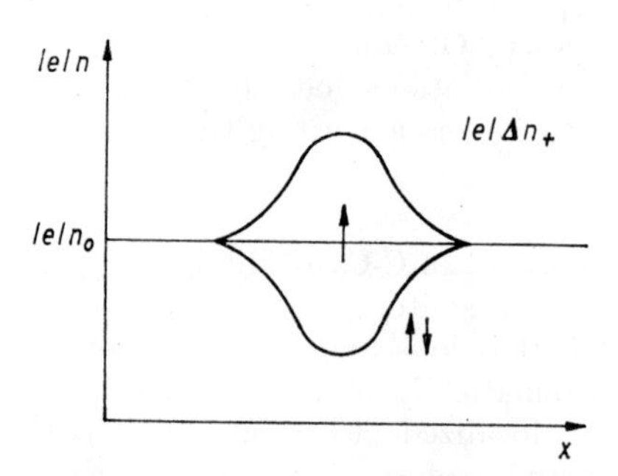

Fig. 6.6-1. Charge distribution of a soliton. The distributions indicated by ↑ and ↑↓, respectively, are caused by the occupation of the soliton state by one or two electrons, respectively. $\varrho_0 = |e|\, n_0$ denotes the original homogeneous charge density

The energies of the localized electron states are within the forbidden gap: the soliton level in the middle of the gap ($\varepsilon = 0$), polaronic levels at $\varepsilon = \pm\Delta/\sqrt{2}$ (class B).

The total energy necessary for the creation of a soliton (lattice distortion plus the change of the electronic continuum) amounts to $E_s = 2\Delta/\pi \approx 0.64\Delta$, that of a polaron $E_p = 2\sqrt{2}\,\Delta/\pi \approx 0.90\Delta$. This means the energy expense for the creation of a soliton or a polaron is smaller than for injecting an electron into the conduction band. Therefore, such injected electrons decay into polarons (or solitons). Because of the interaction of the polarons (solitons), the addition of further electrons (e.g. by doping with donors) leads to the formation of polaron (soliton) bands (Mertsching, Fischbeck 81).

The mechanisms of charge transport are different principally in incommensurate and commensurate systems:

Incommensurate systems:

The motion of a charge density wave is connected with a charge transport. In an ideal system it needs no energy, since the energy of the charge density wave does not depend on the phase Φ. Because of the energy gap in the excitation spectrum the shift of the charge density wave occurs without dissipation of energy (Froehlich mechanism of superconductivity). Defects always present in real systems (pinning centres) prevent the sliding of the charge density waves and, therefore, a non-vanishing d.c. conductivity at zero temperature. It may be expected that high electric fields separate the charge density waves from the pinning centres, which manifests itself in a field dependence of the conductivity. A non-ohmic behaviour of that type has indeed been observed in $NbSe_3$ (Ony, Monceau 77).

Commensurate systems:

The investigation of charge transport mechanisms in those systems has been stimulated considerably by experimental results which revealed that the electrical conductivity of polyacetylene may be increased by doping (e.g. with iodine or with AsF_5) over more than 10 orders of magnitude (Chiang et al. 79). Kivelson (81) explained the charge transport in weakly doped polyacetylene ($y \lesssim 0.03\%$) by phonon-stimulated hopping processes between different localized soliton states. An essential precondition for this transport mechanism is the existence of soliton states which may be recharged. The hopping process occurs essentially in three dimensions (by preference between neighbouring polymer chains). For higher dopant concentrations, all original neutral solitons are charged and the neutral solitons for the hopping process are missing. At still higher concentrations (for $(CH)_x$ near 6%) the transport of charges takes place in polaron or bipolaron bands (class B or C) and in some cases a semiconductor-metal transition may be observed (Kivelson, Heeger 85).

Because of the complex macroscopic morphology of polyacetylene, the interpretation of the experimental results is not yet unequivocal. The temperature dependence of the electric conductivity for different doping concentrations y may be described by (Doege et al. 87)

$$\sigma(T, y) = \sigma_0 \exp \{-[T/T_0(y)]^{-1/2}\} \quad (2)$$

which is known from Mott's theory of variable range hopping (Mott 69). The concentration is contained in (2) only in the scaling temperature $T_0(y)$.

References

Solid State Physics in General

Abrikosov, A. A.: Introduction to the Theory of Normal Metals (Solid State Physics, Suppl. 12). Academic Press, New York 1972

Brauer, W.: Einführung in die Elektronentheorie der Metalle. Geest & Portig, Leipzig 1972

Davydov, A. S.: Theory of Solids (in Russian). Nauka, Moscow 1976
(Давыдов, А. С.: Теория твердых тел. Наука, Москва 1976)

Harrison, W. A.: Electronic Structure and the Properties of Solids. The Physics of the Chemical Bond. W. H. Freeman and Company, San Francisco 1980

Kittel, Ch.: Quantum Theory of Solids. John Wiley & Sons, New York, London 1963

Kittel, Ch.: Introduction to Solid State Physics. John Wiley & Sons, New York 1971

Kreher, K.: Festkörperphysik. Akademie-Verlag, Berlin 1973

Lifschitz, I. M., Asbel, M. J., Kaganow, M. I.: Elektronentheorie der Metalle. Akademie-Verlag, Berlin 1975

Madelung, O.: Festkörpertheorie I, II, III. Springer-Verlag, Berlin, Heidelberg, New York 1972

Schulze, G. E. R.: Metallphysik. Akademie-Verlag, Berlin 1974

Weissmantel, Chr., Hamann, C.: Grundlagen der Festkörperphysik. Deutscher Verlag der Wissenschaften, Berlin 1979

Ziman, J. M.: Principles of the Theory of Solids. Cambridge University Press, London 1972

Recent Monographs and Proceedings

Abrikosov, A. A.: Fundamentals of the Theory of Metals (in Russian). Nauka, Moscow 1987
(Абрикосов, А. А.: Основы теории металлов. Наука Москва 1987)

Bassani, F., Fumi, F., Tosi, M. P. (eds.): Highlights of Condensed-Matter Theory. North-Holland, Amsterdam 1985

Böttger, H.: Principles of the Theory of Lattice Dynamics. Akademie-Verlag, Berlin 1983

Dahl, J. P., Avery, J. (eds.): Local Density Approximation in Quantum Chemistry and Solid State Physics. Plenum Press, New York 1984

Deb, B. M. (ed.): The Force Concept in Chemistry. Van Nostrand Reinhold, New York 1981

Devreese, J. T., Van Camp, P. E. (eds.): Electronic Structure, Dynamics, and Quantum Structural Properties of Condensed Matter. Plenum Press, New York 1985

Devreese, J. T., Van Doren, V. E., Van Camp, P. E. (eds.): Ab initio Calculation of Phonon Spectra. Plenum Press, New York 1985

DREIZLER, R. M., DA PROVIDENCIA, J. (eds.): Density Functional Methods in Physics. Plenum Press, New York 1985

ESCHRIG, H.: Optimized LCAO Method and the Electronic Structure of Extended Systems. Akademie-Verlag, Berlin 1988

FRADIN, F. Y. (ed.): Electronic Structure and Properties (Treatise on Materials Science and Technology, vol. 21). Academic Press, New York 1981

GIRORD, D., ZIEBECK, K. R. A. (eds.): Workshop on 3*d* Metallic Magnetism. ILL, Grenoble 1983

GRIMVALL, G.: Thermophysical Properties of Materials. North-Holland, Amsterdam 1986

HAFNER, J.: From Hamiltonians to Phase Diagrams. The Electronic and Statistical-Mechanical Theory of sp-Bonded Metals and Alloys. Springer-Verlag, Berlin 1987

HILSCHER, G., WIESINGER, G., GRATZ, E., SCHMITZER, C., WEINBERGER, P., GRÖSSINGER, R. (eds.): Proc. Conf. Electronic Structure and Properties of Rare Earth and Actinide Intermetallics. North-Holland, Amsterdam 1985

KRAEFT, W.-D., KREMP, D., EBELING, W., RÖPKE, G.: Quantum Statistics of Charged Particle Systems. Akademie-Verlag, Berlin 1986

KRYACHKO, E. S., LUDEÑA E. V.: Energy Density Functional Theory of Many-Electron Systems. Kluwer Academic Publishers, Dordrecht 1990

LUNDQVIST, S., MARCH, N. H. (eds.): Theory of the Inhomogeneous Electron Gas. Plenum Press, New York 1983

MERTIG, I., MROSAN, E., ZIESCHE, P.: Multiple Scattering Theory of Point Defects in Metals: Electronic Properties, Teubner-Texte zur Physik, vol. 11. Teubner-Verlagsgesellschaft, Leipzig 1987

NEMOSHKALENKO, V. V., ANTONOV, V. N.: Methods of Computational Physics in Solid State Theory. Band Structure Theory of Metals (in Russian). Naukova Dumka, Kiev 1985
(Немошкаленко, В. В., Антонов, В. Н.: Методы вычислительной физики в теории твердых тел. Зонная структура металлов. Наукова Думка, Киев 1985)

NEMOSHKALENKO, V. V., KUCHERENKO, Ju. N.: Methods of Computational Physics in Solid State Theory. Electronic States in Non-ideal Crystals (in Russian). Naukova Dumka, Kiev 1986
(Немошкаленко, В. В., Кучеренко, Ю. Н.: Методы вычислительной физики в теории твердых тел. Электронные состояния в неидеальных кристаллах. Наукова Думка, Киев 1986)

RAO, B. K., JENA, P. (eds.): Physics and Chemistry of Small Clusters. Plenum Press, New York 1987

SPRINGBORG, M. (ed.): Electrons at the Fermi Surface. Cambridge University Press 1980

WEINBERGER, P.: Electron Scattering Theory for Ordered and Disordered Matter. Oxford University Press 1990

WELLER, W., ZIESCHE, P. (eds.): Localization in Disordered Systems. Teubner-Texte zur Physik, vol. 3 and vol. 16. Teubner-Verlagsgesellschaft, Leipzig 1983 and 1986

YUSOUFF, M. (ed.): Electronic Band Structure and Its Applications. Lecture Notes in Physics, vol. 283. Springer-Verlag, Berlin 1987

ZIESCHE, P. (ed.): Proc. 4th—19th Symp. Electronic Structure. Technische Universität Dresden, 1974—1989

ZIESCHE, P., LEHMANN, G., et al.: Ergebnisse in der Elektronentheorie der Metalle. Akademie-Verlag, Berlin 1983; Springer-Verlag, Berlin, Heidelberg, New York, Tokyo 1983

Special Literature Quoted in the Text

AGUILERA-NAVARRO, V. C., BAKER, G. A., Jr., DE LLANO, M.: Phys. Rev. **B 32**, 4502 (1985)

ALLAN, G., LANNOO, M.: Phys. Rev. **B 37**, 2678 (1988)

ANDERSEN, O. K.: Phys. Rev. **B 12**, 3060 (1975)

ANDERSEN, O. K., JEPSEN, O., GLÖTZEL, D.: in BASSANI, F., FUMI, F., TOSI, M. P. (eds.), Highlights of Condensed-Matter Theory. North-Holland, Amsterdam 1985, p. 59

ANDERSEN, O. K., JEPSEN, O., SOB, M.: in YUSOUFF, M. (ed.), Lecture Notes in Physics, vol. 283. Springer-Verlag, Berlin 1987, p. 1

ASADA, T., TERAKURA, K., JARLBORG, T.: J. Phys. **F 11**, 1847 (1981)

ASHCROFT, N. W., MERMIN, N. D.: Solid State Physics. Holt, Rinehart, and Winston, New York 1976

BADRALEXE, E., FREEMAN, A. J.: Phys. Rev. **B 37**, 10469 (1988)

BARTH, U. von, HEDIN, L.: J. Phys. **C 5**, 1629 (1972)

BARTH, U. von, GROSSMANN, G.: Solid State Commun. **32**, 645 (1979)

BECHGAARD, K., CARNEIRO, K., OLSEN, M., RASMUSSEN, F. B., JACOBSEN, C. S.: Phys. Rev. Lett. **46**, 852 (1981)

BEDNORZ, J. G., MUELLER, K. A.: Z. Phys. **B 64**, 189 (1986)

BLÜGEL, S., AKAI, H., ZELLER, R., DEDERICHS, P. H.: Phys. Rev. **B 35**, 3271 (1987)

BORN, M., OPPENHEIMER, R.: Ann. Physik (Leipzig) **84**, 457 (1927)

BRAZOVSKI, S. A.: Ž. eksper. teoret. fiz. **78**, 677 (1980)

BRAZOVSKI, S. A., KIROWA, N. N.: Pisma Ž. eksper. teoret. fiz. **33**, 6 (1981)

BRASPENNING, P. J., ZELLER, R., LODDER, A., DEDERICHS, P. H.: Phys. Rev. **B 29**, 703 (1984)

BREDOV, M. M., KOTOV, B. A., OKUNEVA, N. M., OSKOTSKI, W. S., SHAKH-BUDAGOV, A. L.: Fiz twerd. tela **9**, 289 (1967)

BROOKS, M. S. S.: in V. HEINE (ed.), Europhysics Conference Abstracts **6 A**, 402 (1982); J. Phys. **F 13**, 103 (1983); L 197 (1983)

BROVMAN, E. G., KAGAN, Ju. M.: in G. K. HORTON, A. A. MARADUDIN (ed.), Dynamical Properties of Solids, vol. 1. North Holland, Amsterdam 1974, p. 191

BROWN, R. G.: J. Phys. **B 21**, L 309 (1988)

BUTLER, W. H.: Phys. Rev. **B 15**, 5267 (1977)

BUTLER, W. H., PINSKI, F. J., ALLEN, P. B.: Phys. Rev. **B 19**, 3708 (1979)

BUTLER, W. H.: in F. Y. FRADIN (ed.), Treatise on Materials Science and Technology, vol. 21, Electronic Structure and Properties. Academic Press, New York 1981, p. 165

CHEN, S. P., VOTER, A. F., SROLOVITZ, D. J.: Phys. Rev. Lett. **57**, 1308 (1986)

CHIANG, C. K., HEEGER, A. J., MAC DIARMID, A. G.: Ber. Bunsenges. Phys. Chem. **83**, 407 (1979)

DIMMOCK, J. O.: Solid State Phys. **26**, 103 (1971)

DÖGE, H.-G., ELEFANT, D., WUCKEL, L.: Acta Polymer. **38**, 503 (1987)

DODSON, B. W.: Phys. Rev. Lett. **60**, 2289 (1988)

DUTHIE, J. C., PETTIFOR, D. G.: Phys. Rev. Lett. **38**, 564 (1977)

EASTMAN, D. E., HIMPSEL, F. J., KNAPP, J. A.: Phys. Rev. Lett. **44**, 95 (1980)

EASTMAN, D. E., HIMPSEL, F. J.: in P. RHODES (ed.), Physics of Transition Metals 1980. Inst. Phys. Conf. Ser. **55**, 115 (1981)

EBERT, H., WINTER, H., VOIGTLÄNDER, J.: J. Phys. **F 16**, 1133 (1986)

ECKERN, U., SCHMID, A., WEBER, W., WÜHL, H. (eds.): Proc. LF-17 (parts I and II). North Holland, Amsterdam 1984; part III: Physica **126 B + C**, 1–526 (1984)

EGUILUZ, A. G.: Physica Scr. **36**, 651 (1987)
EHRENREICH, H.: Science **235**, 1029 (1987)
ELIASHBERG, G. M.: Ž. eksper. teoret. fiz. **38**, 966 (1960)
ELLIALTIOGLU, S., ZELLER, R., DEDERICHS, P. H.: J. Phys. **F 17**, 409 (1987)
ERDÖS, P., HERNDON, R. C.: Adv. Phys. **31**, 65 (1982)
ESCHRIG, H., VAN LOYEN, L., ZIESCHE, P.: phys. stat. sol. (b) **66**, 587 (1974)
ESCHRIG, H.; in P. ZIESCHE (ed.), Proc. 6th Symp. Electronic Structure. Technische Universität Dresden 1976, p. 46
ESCHRIG, H., FELDMANN, K., HENNIG, K., MATZ, W., PAUFLER, P.: phys. stat. sol. (b) **79**, 283 (1977)
ESCHRIG, H., RICHTER, R.: in P. ZIESCHE (ed.), Proc. 12th Symp. Electronic Structure. Technische Universität Dresden 1982, p. 7
ESCHRIG, H.: 1983a, s. ZIESCHE, LEHMANN et al. 83, chapter 5
ESCHRIG, H.: 1983b, s. ZIESCHE, LEHMANN et al. 83, chapter 8
ESCHRIG, H.: Optimized LCAO Method and the Electronic Structure of Extended Systems. Akademie-Verlag Berlin 1988; Springer-Verlag, Berlin, Heidelberg, New York 1988
FAULKNER, J. S., STOCKS, G. M.: Phys. Rev. **B 21**, 3222 (1980)
FAULKNER, J. S.: Progr. Mat. Science **27**, 1 (1982)
FAULKNER, J. S., BEAULAC, J. P.: Phys. Rev. **B 26**, 1597 (1982)
FAULKNER, J. S.: Phys. Rev. **B 38**, 1686 (1988)
FRIEDEL, J.: Phil. Mag. **43**, 153 (1952)
FU, C. L., FREEMAN, A. J.: Phys. Rev. **B 37**, 2685 (1988)
FULDE, P.: in P. ZIESCHE (ed.), Proc. 14th Symp. Electronic Structure. Technische Universität Dresden 1984, p. 27
GAY, J. G., RICHTER, R.: Phys. Rev. Lett. **56**, 2728 (1986)
GIES, P.: J. Phys. **C 19**, L 209 (1986)
GIES, P., GERHARDTS, R. R.: Phys. Rev. **B 33**, 982 (1986)
GIORDANO, N.: Phys. Rev. **B 22**, 5635 (1980)
GLÖTZEL, D., RAINER, D., SCHOBER, H. R.: Z. Phys. **B 35**, 317 (1979)
GODBY, R. W., SCHLÜTER, M., SHAM, L. J.: Phys. Rev. **B 35**, 4170 (1987); Phys. Rev. **B 37**, 10159 (1988)
GONIS, A.: Phys. Rev. **B 34**, 8313 (1986)
GONIS, A.: preprint UCRL-97482 (1988)
GONIS, A., STOCKS, G. M., BUTLER, W. H., WINTER, H.: Phys. Rev. **B 29**, 555 (1984)
GRECHUKHIN, D. P., SOLDATOV, A. A.: Ž. eksper. teoret. fiz. **91**, 1969 (1986)
GRIMVALL, G.: The Electron-Phonon Interaction in Metals. North-Holland, Amsterdam 1981
GYORFFY, B. L., KOLLAR, J., PINDOR, A. J., STOCKS, G. M., STAUNTON, J., WINTER, H.: On the Theory of Ferro-Magnetism of Transition Metals at Finite Temperatures. Univ. of Bristol 1984
HAFNER, J.: Phys. Rev. **B 15**, 617 (1977); **B 19**, 5094 (1979)
HAFNER, J., HEINE, V.: J. Phys. **F 13**, 2479 (1983)
HAFNER, J.: From Hamiltonians to Phase Diagrams. Springer-Verlag, Berlin 1987, pp. 323
HAGEMANN, H. J., GUDAT, W., KUNZ, C.: J. Opt. Soc. Am. **65**, 742, (1975)
HAMANN, C., HÖHNE, H. J., KERSTEN, F., MÜLLER, M., PRZYBOROWSKI, F., STARKE, M.: phys. stat. sol. (a) **50**, K 189 (1978)
HAMANN, D. R., SCHLÜTER, M., CHIANG, C.: Phys. Rev. Lett. **43**, 1494 (1979)
HARRISON, W. A.: Pseudopotentials in the Theory of Metals. Benjamin, New York 1966

HAYDOCK, R., JOHANNES, P. L.: J. Phys. **F 5**, 2055 (1975)
HEDIN, L., LUNDQVIST, S.: Solid State Physics **23**, 1 (1969)
HEDIN, L., LUNDQVIST, B. I.: J. Phys. **C 4**, 2064 (1971)
HEEGER, A. J., MAC DIARMID, A. G.: in J. BERNASCONI, T. SCHNEIDER (eds.), Physics in One Dimension. Springer-Verlag, Berlin 1981, p. 179
HEINE, V., ABARENKOV, I. V.: Phil. Mag. **9**, 451 (1964)
HEINE, V.: in J. M. ZIMAN (ed.), The Physics of Metals, 1. Electrons. Cambridge University Press, 1969, p. 41
HERRING, C.: in Magnetism, vol. IV, G. T. RADO, H. SUHL (eds.), Academic Press, New York 1966
HO, K.-M., BOHNEN, K. P.: Phys. Rev. **B 32**, 3446 (1985)
HO, K.-M., BOHNEN, K. P.: Phys. Rev. Lett. **59**, 1833 (1987)
HOHENBERG, P., KOHN, W.: Phys. Rev. **136**, B 864 (1964)
HOPFIELD, J. J.: Phys. Rev. **186**, 443 (1969)
HORSCH, P., VON DER LINDEN, W., LUKAS, W.-D.: Solid State Commun. **62**, 359 (1987)
HUBBARD, J.: J. Phys. **C 2**, 1222 (1969)
HYBERTSEN, M. S., LOUIE S. G.: Phys. Rev. **B 35**, 5585 (1987)
IHM, J., ZUNGER, A., COHEN, M. L.: J. Phys. **C 12**, 4409 (1979)
INGLESFIELD, J. E.: Prog. Surface Sci. **20**, 105 (1985)
JENA, P.: in F. Y. FRADIN (ed.), Treatise on Materials Science and Technology, vol. 21, Electronic Structure and Properties. Academic Press, New York 1981, p. 351
JEPSEN, O.: Phys. Rev. **B 12**, 2988 (1975)
JEPSEN, O., MADSEN, J., ANDERSON, O. K.: Phys. Rev. **B 26**, 2790 (1982)
JIANG, P., MARCUS, P. M., JONA, F.: Solid State Commun. **59**, 275 (1986)
JOHN, W.: 1983, s. ZIESCHE, LEHMANN et al. 83, chapter 9
JOHN, W., HAMANN, D., URWANK, P.: phys. stat. sol. (b) **86**, 596 (1978)
JOHN, W., NEMOSHKALENKO, V. V., ANTONOV, V. N., ANTONOV, Vl. N.: in P. ZIESCHE (ed.), Proc. 12th Symp. Electronic Structure. Technische Universität Dresden 1982, p. 244
KAGAN, JU. M.: Ž. eksper. teoret. fiz. **42**, 1375 (1962)
KAGAN, JU. M., PUSHKAROV, W. W., HOLAS, A.: Ž. eksper. teoret. fiz. **73**, 967 (1977)
KASCHNER, R., ZIESCHE, P.: Physica Scr. **38**, 414 (1988a)
KASCHNER, R., ZIESCHE, P.: preprint, Nordita-88/28 S (1988b)
KERKER, G. P.: J. Phys. **C 13**, L 189 (1980)
KHANNA, S. N., RAO, B. K., JENA, P., ESTERLING, D., PUSKA, M. J.: Phys. Rev. **B 37**, 6 (1988)
KIVELSON, S.: Phys. Rev. Lett. **46**, 1344 (1981)
KIVELSON, S., HEEGER, A. J.: Phys. Rev. Lett. **55**, 308 (1985)
KOELLING, D. D., ARBMANN, G. O.: J. Phys. **F 5**, 2041 (1975)
KOHN, W., ROSTOKER, N.: Phys. Rev. **94**, 1111 (1954)
KOHN, W., SHAM, L. J.: Phys. Rev. **140**, A 1133 (1965)
KORRINGA, J.: Physica (Utrecht) **13**, 392 (1947)
KÜBLER, J.: J. Magn. Magn. Mater. **20**, 277 (1980)
LANDAU, L. D., LIFSCHITZ, E. M.: Lehrbuch der Theoretischen Physik, vol. V, Statistische Physik, part 1. Akademie-Verlag, Berlin 1987, § 163
LANDAU, L. D., LIFSCHITZ, E. M.: Lehrbuch der Theoretischen Physik, vol. IX, Statistische Physik, part 2. Akademie-Verlag, Berlin 1984
LANG, N. D., KOHN, W.: Phys. Rev. **B 1**, 4555 (1970); **B 3**, 1215 (1971)
LEARY, K. J., ZUR LOYE, H. C., KELLER, S. W., FALTENS, T. A., HAM, W. K., MICHAELS, J. N., STADY, A. M.: Phys. Rev. Lett. **59**, 1236 (1987)

LEE, J. I., FU, C. L., FREEMAN, A. J.: Phys. Rev. **B 36**, 9318 (1987)

LEHMANN, G., RENNERT, P., TAUT, M., WONN, H.: phys. stat. sol. **37**, K 27 (1970)

LEHMANN, G., PEGEL, B.: Elektronenstruktur und mechanische Eigenschaften. Deutscher Verlag für Grundstoffindustrie, Leipzig 1972, p. 191

LEHMANN, G.: Ann. Physik (Leipzig) **30**, 155 (1973b)

LEHMANN, G.: phys. stat. sol. (b) **56**, K 33 (1973a)

LEHMANN, G.: Kristall und Technik **9**, 587 (1974)

LEHMANN, G.: phys. stat. sol. (b) **70**, 737 (1975)

LEHMANN, G.: 1983, s. ZIESCHE, P., LEHMANN, G. et al. 83, chapter 4

LENK, R.: Theorie elektromagnetischer Felder. Deutscher Verlag der Wissenschaften, Berlin 1976

LOUCKS, T. L.: Augmented Plane Wave Method. Benjamin, New York 1967

LUCAS, A. A.: in B. DI BARTOLO, J. DANKO (eds.), Collective Excitations in Solids. Plenum Press, New York 1983, pp. 365

LUO, J. S., LEGRAND, B.: Phys. Rev. **B 38**, 1728 (1988)

MACKINTOSH, A. R., ANDERSEN, O. K.: in M. SPRINGFORD (ed.), Electrons at the Fermi Surface. Cambridge University Press, 1980, p. 149

MAHAN, D. G.: Phys. Rev. **B 21**, 1421 (1980)

MAKSIMOV, E. G., KHOMSKI, D. I., in: GINSBURG, V. L. KIRZHNITS (eds.), Problems of High-temperature Superconductivity, Nauka, Moscow 1977
(Максимов, Е. Г., Хомский, Д. И.: Проблема высокотемпературной сверхпроводимости, под редакцией В. Л. Гинзбурга и Д. А. Киржница, Наука, Москва 1977)

MATTHEISS, L. F.: Phys. Rev. **B 1**, 373 (1970)

MCMAHAN, A. K., MORIARTY, J. A.: Phys. Rev. **B 27**, 3235 (1983)

MCMILLAN, W. L.: Phys. Rev. **167**, 331 (1968)

MEARNS, D.: Phys. Rev. **B 38**, 5906 (1988)

MERTIG, I.: Dissertation A, Technische Universität Dresden, 1982

MERTIG, I., MROSAN, E., SCHÖPKE, R.: J. Phys. **F 12**, 1689 (1982)

MERTIG, I., MROSAN, E.: J. Phys. **F 12**, 3031 (1982); **F 3**, 373 (1983)

MERTIG, I., MROSAN, E., ZIESCHE, P.: Multiple Scattering Theory of Point Defects in Metals: Electronic Properties (Teubner-Texte zur Physik, vol. 11). Teubner Verlagsgesellschaft, Leipzig 1987

MERTSCHING, J., FISCHBECK, H. J.: phys. stat. sol. (b) **103**, 783 (1981)

MÖBIUS, A., GOEDSCHE, F., VOJTA, G.: Physica **95 A**, 294 (1979)

MOLENAAR, J.: J. Phys. **C 21**, 1455 (1988); RANA 88-18, Eindhoven University of Technology (1988). J. Phys. Cond. Mat. **1**, 6559 (1989)

MORUZZI, V. L., WILLIAMS, A. R., JANAK, J. F.: Phys. Rev. **B 15**, 2854 (1977)

MORUZZI, V. L., JANAK, J. L., WILLIAMS, A. R.: Calculated Electronic Properties of Metals. Pergamon Press, New York 1978

MORUZZI, V. L., MARCUS, P. M., SCHWARZ, K. MOHN, P.: Phys. Rev. **B 34**, 1784 (1986)

MORUZZI, V. L., MARCUS, P. M.: Phys. Rev. **B 38**, 1613 (1988)

MOTT, N. F.: Phil. Mag. **19**, 835 (1969)

MOTT, N. F.: Metal Insulator Transitions. Taylor & Francis, London 1974

MROSAN, E., LEHMANN, G.: phys. stat. sol. (b) **77**, 607 (1976); **78**, 159 (1976)

MROSAN, E., LEHMANN, G.: phys. stat. sol. (b) **87**, K 21 (1978)

MROSAN, E., LEHMANN, G.: 1983, s. ZIESCHE, LEHMANN et al. 83, chapter 11

MÜLLER, Ch., WONN, H., BLAU, W., KRIVITZKIJ, V. P.: in P. ZIESCHE (ed.), Proc. 8th Symp. Electronic Structure. Technische Universität Dresden 1978, p. 189

MUKHERJEE, S., MORAN-LOPEZ, J. L., KUMAR, V., BENNEMANN, K. H.: Phys. Rev. **B 25**, 730 (1982)

NEEDS, R. J.: Phys. Rev. Lett. **58**, 53 (1987)

NEMOSHKALENKO, V. V., ANTONOV, V. N., ANTONOV, Vl. N., JOHH, W., WONN, H., ZIESCHE, P.: phys. stat. sol. (b) **111**, 11 (1982)

NIELSEN, O. H., MARTIN, R. M.: Phys. Rev. Lett. **50**, 697 (1983); Phys. Rev. **B 32**, 3780 (1985); **B 35**, 9308 E (1987)

NIELSEN, O. H., MARTIN, R. M.: Phys. Rev. **B 32**, 3792 (1985)

NOSKOV, M. M., Optical and Magneto-optical Properties of Metals (in Russian). Ac. Sc. USSR, Ural Scient. Centre, Sverdlovsk, 1983, p. 75

OGUCHI, T., FREEMAN, A. J.: J. Magn. Magn. Mater. **46**, L 1—L 4 (1986)

OLÉS, A., STOLLHOFF, G.: Phys. Rev. **B 29**, 314 (1984)

ONY, N. P., MONCEAU, P.: Phys. Rev. **B 16**, 3443 (1977)

OSSICINI, S., BERTONI, C. M., GIES, P.: Surface Sci. **178**, 244 (1986)

OSSICINI, S., FINOCCHI, F., BERTONI, C. M.: Surface Sci. **189/190**, 776 (1987)

PAASCH, G.: Dissertation A, Technische Universität Dresden 1970 and phys. stat. sol. **38**, K 123 (1970)

PAASCH, G., WOITTENNEK, H.: phys. stat. sol. (b) **55**, 537 (1973)

PAASCH, G., HIETSCHOLD, M.: phys. stat. sol. (b) **83**, 209 (1977)

PAASCH, G., HIETSCHOLD, M.: 1983, s. ZIESCHE, LEHMANN et al. 83, chapter 10

PAPACONSTANTOPOULOS, A. A., BOYER, L. L., KLEIN, B. M., WILLIAMS, A. R., MORUZZI, V. L., JANAK, J. F.: Phys. Rev. **B 15**, 4221 (1977)

PAUFLER, P., LEUSCHNER, D.: Kristallografische Grundbegriffe der Festkörperphysik. Akademie-Verlag, Berlin 1975

PAYNE, M. C.: J. Phys. **C 20**, L 983 (1987)

PERDEW, J. P., ZUNGER, A.: Phys. Rev. **B 23**, 5048 (1981)

PERROT, F., RASOLT, M.: Phys. Rev. **B 37**, 2147 (1988)

PETTIFOR, D. G.: J. Phys. **C 2**, 1051 (1969)

PETTIFOR, D. G.: J. Phys. **C 3**, 367 (1970)

PETTIFOR, D. G.: J. Phys. **C 5**, 97 (1972)

PETTIFOR, D. G.: J. Phys. **F 7**, 1009 (1977)

PETTIFOR, D. G.: New Scientist, 29 May 1986, p. 48

PINSKI, F. J., ALLEN, P. B., BUTLER, W. H.: Phys. Rev. **B 23**, 5080 (1981)

PODLOUCKY, R., ZELLER, R., DEDERICHS, P. H.: Phys. Rev. **B 22**, 5777 (1980)

POSTERNAK, M., KRAKAUER, H., FREEMAN, A. J.: Phys. Rev. **B 25**, 755 (1982)

QIAN, G., CHADI, D. J.: Phys. Rev. **B 35**, 1288 (1987)

RADWAN, A.: Dissertation A, Technische Universität Dresden 1976

REDINGER, J., WEINBERGER, P., NECKEL, A.: Phys. Rev. **B 35**, 5647 (1987)

RENNERT, P.: 1983, s. ZIESCHE, LEHMANN et al. 83, chapter 7

RENNET, P., LEHMANN, G.: in Intermetallische Phasen. Deutscher Verlag für Grundstoffindustrie, Leipzig 1977, p. 91

RENNERT, P., PAASCH, G.: 1983, s. ZIESCHE, LEHMANN et al. 83, chapter 2

RENNERT, P., HUNG, N. V.: phys. stat. sol. (b) **148**, 49 (1988)

RICHTER, R., ESCHRIG, H.: Proc. Conf. EPS Condensed Matter Division, Pisa, 1987

RICHTER, R., ESCHRIG, H., VELICKY, B.: J. Phys. **F 17**, 351 (1987)

RÖPKE, G., ZIESCHE, P.: 75 Jahre Quantentheorie, Abh. d. AdW d. DDR, Abt. Math., Nat. wiss., Technik, 7 N. Akademie-Verlag, Berlin 1977, p. 269

SCHIMMEL, Th., RIESS, W., GMEINER, J., DENNINGER, G., SCHWOERER, M., NAARMANN, H., THEOPHILOU, N.: Solid State Commun. **65**, 1311 (1988)

SCHOBER, C., KURZ, G., WONN. H., NEMOSHKALENKO, V. V., ANTONOV, V. N.: phys. stat. sol. **136**, 233 (1986)

SHAW jr., R. W.: Phys. Rev. **174**, 769 (1968)

SHENOY, G. K., WAGNER, F. E. (eds.): Mößbauer Isomer Shifts. North-Holland, Amsterdam 1978

SHILES, E., SASAKI, T., INOKUTI, M., SMITH, D. Y.: Phys. Rev. **B 22**, 1612 (1980)

SIEGBAHN, K., et al.: ESCA, Atomic, Molecular and Solid State Structure Studied by Means of Electron Spectroscopy. Almqvist and Wiksells, Uppsala 1967

SINGWI, K. S., SJÖLANDER, A., TOSI, A. P., LAND, R. M.: Phys. Rev. **B 1**, 1044 (1970)

SKRIVER, H. L.: J. Phys. **F 11**, 97 (1981)

SKRIVER, H. L.: Phys. Rev. Lett. **49**, 1768 (1982)

SKRIVER, H. L.: The Linear Muffin-Tin Orbital Method. Springer-Verlag, Berlin 1984

SKRIVER, H. L., JOHANSSON, B., ANDERSEN, O. K.: in V. HEINE (ed.), Europhysics Conference Abstracts **6 A**, 44 (1982); also in J. FREEMAN, G. H. LÄNDER, J. KELLER (eds.): Handbook on the Physics and Chemistry of the Actinides, 1984

SKRIVER, H. L., MERTIG, I.: Phys. Rev. **B 15**, 4431 (1985)

SKRIVER, H. L., ERIKSSON, O., MERTIG, I., MROSAN, E.: Phys. Rev. **B**, **37**, 1706 (1988)

SLATER, J. C.: Phys. Rev. **51**, 846 (1937)

SLATER, J. C.: Phys. Rev. **81**, 385 (1951); **82**, 538 (1951)

SLATER, J. C., KOSTER, G. F.: Phys. Rev. **94**, 1498 (1954)

SLATER, J. C.: The Self-Consistent Field for Molecules and Solids. Mc Graw-Hill, New York 1974

SMITH, J. R.: Phys. Rev. **181**, 522 (1969)

SMRCKA, L.: in P. ZIESCHE, (ed.), Proc. 12th Symp. Electronic Structure. Technische Universität Dresden 1982

SOKOLOV, J., JONA, F., MARCUS, P. M.: Phys. Rev. **B 33**, 1397 (1986)

SOVEN, P.: Phys. Rev. **156**, 809 (1967)

STEFANOU, N., OSWALD, A., ZELLER, R., DEDERICHS, P. H.: Phys. Rev. **B 35**, 6911 (1987)

SU, W. P., SCHRIEFFER, J. R., HEEGER, A. J.: Phys. Rev. Lett. **42**, 1698 (1979), Phys. Rev. **B 22**, 2099 (1980)

TACHIKI, M.: Helv. Phys. Acta **56**, 189 (1983)

TAUT, M.: 1983, s. ZIESCHE, LEHMANN et al. 83, chapter 6

TAUT, M., HANKE, W.: phys. stat. sol. (b) **77**, 543 (1976)

THOULESS, D. J.: Phys. Rev. Lett. **39**, 1167 (1977)

URWANK, P.: Dissertation A, Technische Universität Dresden 1974

VANDERBILT, D.: Phys. Rev. Lett. **59**, 1456 (1987)

VELICKY, B., KIRKPATRICK, S., EHRENREICH, H.: Phys. Rev. **175**, 747 (1968)

VONSOVSKI, S. V., IZJUMOV, JU. A., KURMAEV, E. Z.: Superconductivity of Transition Metals, Their Alloys and Compounds. Nauka, Moscow 1977
(Вонзовский, С. В., Изюмов, Ю. А., Курмаев, Э. З.: Сверхпроводимость переходных металлов, их сплавов и соединений. Наука, Москва 1977)

VOSKO, S. H., WILK, L., NUSAIR, M.: Can. J. Phys. **58**, 1200 (1986)

WEBER, W.: in P. PHARISEAU, W. TEMMERMANN (eds.), Electronic Structure of Complex Systems. Plenum Press, New York 1984

WEINBERGER, P.: Electron Scattering Theory for Ordered and Disordered Matter, Oxford University Press 1990

WEINBERGER, P., GONIS, A.: in A. J. FREEMAN, G. H. LANDER (eds.), Handbook on the Physics and Chemistry of the Actinides. Elsevier Science Publ., Amsterdam 1987, p. 1.

WEINBERGER, P., DIRL, R., BORING, A. M., GONIS, A., FREEMAN, A. J.: Phys. Rev. **B 37**, 1383 (1988)
WELLER, W., ZIESCHE, P.: Localization in Disordered Systems (Teubner-Texte zur Physik, vols. 3 and 16). Teubner Verlagsgesellschaft, Leipzig 1984 and 1987
WETTE, DE, F. W., KRESS, W., SCHRÖDER, U.: Phys. Rev. **B 32**, 4143 (1985)
WINTER, J.: Magnetic Resonance in Metals. University Press, Oxford 1971
WOITTENNEK, H.: Dissertation A, Technische Universität Dresden 1975
WU, M. K., ASHBURN, J. R., TORNG, C. J., HOR, P. H., MENG, R. L., GAO, L., HUANG, Z. I., WANG, Y. O., CHU, C. W.: Phys. Rev. Lett. **58**, 908 (1987)
WU, T. C., VOGEL, F. L., PENDRY, L. A., ZELLER, C.: Mat. Science and Engineering **47**, 161 (1981)
YUSOUFF, M. (ed.): Electronic Band Structure and its Applications. Springer-Verlag, Berlin 1987
ZELLER, R., DEDERICHS, P. H.: Phys. Rev. Lett. **42**, 1713 (1979)
ZELLER, R., PODLOUCKY, R., DEDERICHS, P. H.: Z. Phys. **B 83**, 165 (1980)
ZELLER, R.: in M. YUSOUFF (ed.), Current Trends in the Physics of Materials. World Sc. Publ., Singapore 1987a, p. 332
ZELLER, R.: J. Phys. **F 17**, 2123 (1987b)
ZELLER, R.: in M. YUSOUFF (ed.), Lecture Notes in Physics, vol. 283, Springer-Verlag, Berlin 1987c, p. 106
ZELLER, R.: Phys. Rev. **B 38**, 5993 (1988)
ZIESCHE, P.: J. Phys. **C 7**, 1085 (1974)
ZIESCHE, P.: in Intermetallische Phasen. Deutscher Verlag für Grundstoffindustrie, Leipzig 1977, p. 115
ZIESCHE, P., WONN, H., MÜLLER, Ch., NEMOSHKALENKO, V. V., KRIVITZKIJ, V. P.: phys. stat. sol. (b) **87**, 129 (1978)
ZIESCHE, P., LEHMANN, G., et al.: Ergebnisse in der Elektronentheorie der Metalle. Akademie-Verlag, Berlin 1983; Springer-Verlag, Berlin, Heidelberg, New York, Tokyo 1983
ZIESCHE, P., LEHMANN, G., ESCHRIG, H.: 1983, s. ZIESCHE, LEHMANN et al. 83, chapter 0
ZIESCHE, P., PAASCH, G.: 1983, s. ZIESCHE, LEHMANN et al. 83, chapter 1
ZIESCHE, P.: Ann. Physik **45**, 626 (1988)
ZIESCHE, P., GRÄFENSTEIN, J., NIELSEN, O. H.: Phys. Rev. **B 37**, 8167 (1988)
ZIMAN, J. M.: Solid State Physics **26**, 1 (1971)
ZIMAN, J. M.: Electrons and Phonons. Clarendon, Oxford 1976

Summaries

Chapter 1

Nearly freely moving conduction electrons determine the essential properties of metals. They form a degenerate Fermi gas. The periodicity of the lattice causes the appearance of allowed and forbidden energy regions for the electrons. For the calculation of the properties of a solid, the band structure within the Brillouin zone must be known. In the ground state of the total system, the occupied one-particle states in k-space define the Fermi body.

The adiabatic approximation allows separation of the motion of the electrons and the lattice, i.e. of the phonons. The one-particle picture of metallic electrons may be deduced approximately from the many-body problem of interacting electrons. The local density approximation proved to be particularly suitable and successful in calculating the one-particle crystal potential.

Chapter 2

Various methods of band structure calculation exist. In simple metals, the effective crystal potential is weak only. Therefore, the pseudopotential method is well suited for calculation of the band structure. For more general cases, the tight-binding, APW and KKR methods are available. The latter use a special approximation for the lattice potential — the muffin-tin potential — which permits calculation of the band structure via multiple scattering of electrons on the potential wells of the single atoms. Recently, linearized methods have been favoured increasingly because of their possibility to shorten computing time. The main idea of the coherent potential well suited for describing disordered alloys is represented briefly.

Chapter 3

Simple metals with conduction bands originating from atomic s- and p-levels show a weak effective crystal potential. Therefore, their band structure can be understood starting from nearly free electrons. The transition metals are characterized by spatially more localized d-states causing narrow d-bands intersecting the bands of nearly free electrons. The essential parameters are the position and width of the d-bands, and the position of the Fermi energy which is determined by the number of available valence electrons. As a function of the given lattice structure, the densities of states show a characteristic structure.

Because of their larger elementary cells, intermetallic compounds have more conduction bands. The observed lattice structures are also more diverse than for pure metals. The results for CsCl structure are discussed in more detail. The peculiarities of disordered alloys are explained briefly.

Chapter 4

Under the influence of external fields, in particular electrons near the Fermi energy are excited. The dynamic behaviour of electrons is determined by the band structure. Introducing the effective mass, it can be treated in a semi-classical way. The study of the electronic reaction on external fields under the presence of a constant magnetic field is well suited for the determination of the Fermi surface and the band structure near the Fermi energy. Photoelectric and X-ray spectroscopy also yield information on energy regions remote from the Fermi energy and partially on the dispersion relation of the electrons.

Chapter 5

Of course, the band structure $\varepsilon_{\boldsymbol{k},\nu}$ and corresponding Bloch wave functions $\psi_{\boldsymbol{k},\nu}(\boldsymbol{r})$ together with the phonon dispersion $\omega_{\boldsymbol{q},\lambda}$ and the corresponding polarization vectors $\boldsymbol{e}_{\boldsymbol{q},\lambda}$ are finally responsible for all properties of a metal. However, some properties (such as the behaviour of the dislocation ensemble, which determines the plastic properties) are influenced by the above-mentioned basic quantities only very indirectly over several intermediate steps (electronic structures of defects, dislocations, Peierls force, interaction between dislocations). On the other hand, there are some metal properties that are immediately determined by the above input properties. These are discussed in the following in addition to [4].

Chapter 6

The physics of surfaces and interfaces takes an increased practical meaning. Only two-dimensional symmetry is seen, which is the cause of some new effects such as surface states. For the example of pure surfaces of simple metals, typical quantities such as electronic density, dipole barrier, work function, lattice relaxation, and surface energy are discussed qualitatively and quantitatively.

In systems with one- or two-dimensional structures, the binding strength in some direction or within a plane dominates. With the reduced dimension of a physical system, new phenomena appear. In the following, the physical content of the Peierls instability, the enhanced electron-phonon interaction, and the resulting partly unusual consequences for the low-energy excitation spectrum of these systems are discussed.

Index of Abbreviations

APW	augmented plane wave	2.4
ARPES	angle resolved PES	4.3.3
ARXPS	angle resolved XPS	4.3.3
ASA	atomic sphere approximation	2.6.3
bcc	body centered cubic	3.2
BCS	Bardeen-Cooper-Schrieffer (theory)	5.1.3
BZ	Brillouin zone	1.2.2
CPA	coherent potential approximation	2.7
dHvA	de Haas—van Alphen (effect)	4.2.3
DLM	disordered local moment	3.2
DOS	density of states	1.2.2
ESCA	electronic spectroscopy for chemical analysis	4.3.3
fcc	face centered cubic	3.2
GIC	graphite intercalation compound	6.3.2
hcp	hexagonal close packed	3.2
H-NFE-TB	hybridized-nearly free electron-tight binding	2.6.2
INS	inelastic neutron scattering	5.3.3
KCP	potassium cyanoplatinate	6.3.2
KKR	Korringa-Kohn-Rostoker (method)	2.5.1
KSG	Kohn-Sham-Gaspar (exchange)	1.3.2
LAPW	linear APW	2.6.3
LBT	local band theory	3.2
LCAO	linear combination of atomic orbitals	2.1.1
LDA	local density approximation	1.3.2
LEED	low-energy electron diffraction	6
LFC	local field correction	5.2.3
LKKR	linear KKR (method)	2.6.2
LMTO	linear muffin-tin orbital	2.6.3
MT	muffin-tin (potential)	2.3.1
NFE	nearly free electron	2.2.1
OPW	orthogonalized plane wave	2.2.1
PES	photo-electron spectroscopy	4.3.3
RHEED	reflective high-energy diffraction	6
RKKR	relativistic KKR method	2.7
RMTA	rigid MT approximation	5.4.3
RPA	random phase approximation	5.2.3
SSA	single site approximation	2.7
TB	thight binding	2.1.1
TCNQ	tetracyanoquinodimethane	6.3.2

Fundamental Constants and Atomic Units

$\hbar = 1{,}055 \cdot 10^{-34}\,\mathrm{Ws}^2$	Planck's quantum of action
$m = 9{,}110 \cdot 10^{-31}\,\mathrm{kg}$	electron rest mass
$e = -1{,}602 \cdot 10^{-19}\,\mathrm{As}$	charge of electron
$\in^2 = \dfrac{e^2}{4\pi\varepsilon_0}$	coupling constant of Coulomb interaction ($\in^2/r$, $4\pi \in^2/q^2$)
$k_\mathrm{B} = 1{,}381 \cdot 10^{-23}\,\mathrm{Ws/K}$	Boltzmann's constant
$a_0 = \dfrac{\hbar^2}{m\in^2} = 0{,}5292 \cdot 10^{-10}\,\mathrm{m}$	Bohr radius (atomic length unit)
$ryd = \dfrac{\hbar^2}{2ma_0^2} = \dfrac{\in^2}{2a_0} = 13{,}60535\,\mathrm{eV}$	Rydberg (atomic energy unit)

Characteristic Quantities of the Electron Gas

$n = N/V$	density
$r_s = \sqrt[3]{3/4\pi n}/a_0$	dimensionless density parameter
$k_\mathrm{F} = \sqrt[3]{3\pi^2 n}$	Fermi wave number
$\varepsilon_\mathrm{F}^0 = (\hbar k_\mathrm{F})^2/2\,\mathrm{m}$	Fermi energy
$D_\mathrm{F}^0 = 3/2\varepsilon_\mathrm{F}^0$	Fermi density of states
$q_\mathrm{TF} = \sqrt{D_\mathrm{F}^0 4\pi \in^2 n}$	Thomas-Fermi wave number
$\omega_\mathrm{pl} = \sqrt{4\pi \in^2 n/m}$	plasma frequency
$v_q = 4\pi \in^2/q^2$	Fourier transform of Coulomb interaction

Subject Index

Continued from page 6